ISAAC ASIMOV is undoubtedly America's foremost writer on science for the layman. An Associate Professor of Biochemistry at the Boston University School of Medicine, he has written well over a hundred books, as well as hundreds of articles in publications ranging from *Esquire* to Atomic Energy Commission pamphlets. Famed for his science fiction writing (his three-volume Hugo Award-winning THE FOUNDATION TRILOGY is available in individual Avon editions and as a one-volume Equinox edition), Dr. Asimov is equally acclaimed for such standards of science reportage as THE UNIVERSE, LIFE AND ENERGY, THE SOLAR SYSTEM AND BACK, ASIMOV'S BIOGRAPHICAL ENCYCLOPEDIA OF SCIENCE AND TECHNOLOGY, and ADDING A DIMENSION (all available in Avon editions). His non-science writings include the two-volume ASIMOV'S GUIDE TO SHAKESPEARE, ASIMOV'S ANNOTATED DON JUAN, and the two-volume ASIMOV'S GUIDE TO THE BIBLE (available in a two-volume Avon edition). Born in Russia, Asimov came to this country with his parents at the age of three, and grew up in Brooklyn. In 1948 he received his Ph.D. in Chemistry at Columbia and then joined the faculty at Boston University, where he works today.

Avon Books by
Isaac Asimov

The Foundation Trilogy	20933	$3.95
Asimov's Guide to the Bible, the New Testament	24788	$4.95
Asimov's Guide to the Bible, the Old Testament	24794	$4.95
Fact and Fancy	10306	$1.25
From Earth to Heaven	10421	$1.25
Life and Energy	23135	$1.50
Solar System and Back	10157	$1.25
Universe from Flat Earth to Quasar	19828	$1.25
Foundation	23168	$1.25
Foundation and Empire	23176	$1.25
Second Foundation	06429	$ .95
Adding a Dimension	22673	$1.25
Of Time and Space and Other Things	24166	$1.50

THE NEUTRINO

Ghost Particle of the Atom

ISAAC ASIMOV

 DISCUS BOOKS/PUBLISHED BY AVON

Illustrated by Howard S. Friedman

AVON BOOKS
A division of
The Hearst Corporation
959 Eighth Avenue
New York, New York 10019

Library of Congress Catalog Card Number: 66-17073

ISBN: 0-380-00483-6

First Discus Printing, September, 1975

Printed in the U.S.A.

To Walter I. Bradbury,
with great gladness.

CONTENTS

INTRODUCTION
The Riddle of the Universe   xi

1 MOMENTUM   1
GENERALIZATIONS   1
BOUNCING BILLIARD BALLS   4
CONSERVATION OF MOMENTUM   10
CONSERVATION OF ANGULAR MOMENTUM   16

2 ENERGY   20
CONSERVATION OF MASS   20
CONSERVATION OF ENERGY   22
UNIVERSAL GENERALIZATION   27
THE SUN'S ENERGY   32

3 ATOMIC STRUCTURE   39
RADIOACTIVITY   39
THE ATOMIC NUCLEUS   41
NUCLEAR ENERGY   45

4 MASS-ENERGY   49
THE NONCONSERVATION OF MASS   49
RELATIVITY   53
CONSERVATION OF MASS-ENERGY   56
PHOTONS   58

5 ELECTRIC CHARGE 64
   CONSERVATION OF ELECTRIC CHARGE 64
   NUCLEAR REACTIONS AND ELECTRIC CHARGE 66
   NUCLEAR STRUCTURE 69
   NEUTRON BREAKDOWN 76

6 ANTIPARTICLES 81
   LEPTONS AND BARYONS 81
   POSITRONS 83
   ANTINUCLEONS 88
   CONSERVATION OF BARYON NUMBER 91

7 ENTER THE NEUTRINO 97
   ALPHA PARTICLES AND ENERGY 97
   BETA PARTICLES AND ENERGY 100
   THE NECESSITY OF THE NEUTRINO 104
   CONSERVATION OF LEPTON NUMBER 107

8 DETECTION OF NEUTRINOS 110
   THE ABSORPTION OF PHOTONS 110
   THE ABSORPTION OF NEUTRINOS 113
   TRAPPING THE ANTINEUTRINO 116

9 NEUTRINO ASTRONOMY 122
   ANTINEUTRINOS AND THE EARTH 122
   NEUTRINOS AND THE SUN 124
   TRAPPING THE NEUTRINO 127
   SUPERNOVAS AND NEUTRINOS 130
   THE UNIVERSE AND NEUTRINOS 135

10 THE NUCLEAR FIELD 138
   REPULSION WITHIN THE NUCLEUS 138

ATTRACTION WITHIN THE NUCLEUS   141

THE UNCERTAINTY PRINCIPLE   144

UNCERTAINTY AND CONSERVATION   149

11 MUONS   153

TRAPPING THE MESON   153

STRONG AND WEAK INTERACTIONS   156

THE MASSIVE ELECTRON   158

12 THE MUON-NEUTRINO   163

PION BREAKDOWN   163

CONSERVATION OF ELECTRON FAMILY
    NUMBER AND OF MUON FAMILY
    NUMBER   165

THE TWO-NEUTRINO EXPERIMENT   170

CONSERVATION OF PARITY   172

THE GRAVITON   177

APPENDIX
Exponential Numbers   180

INDEX   183

Introduction

THE RIDDLE OF THE UNIVERSE

Probably as long as manlike creatures have found time to withdraw their attention from mere survival and to lift their eyes to the universe about them, they have been impelled to wonder.

What is the universe made of? What governs its workings? What is its purpose?

The questions are endless; the attempts at answers uncertain.

The first groping answers of which we still have record are grouped by us under the heading of "myth." Then, twenty-five centuries ago, came the Greeks, among whom arose the first systematic attempts to apply reason, rather than mystic insight, to solving the riddle of the universe.

It was not until the late sixteenth century—less than four hundred years ago—that scientists came to appreciate fully that reason alone is not enough, either. Experimentation must be added.

The rise of experimental science from 1580 onward has supplied mankind with part answers to the riddle of the universe; part answers of a depth and scope he could not have imagined earlier. Man's mind soared outward to a cosmos inconceivably large and drove inward to tiny structures inconceivably small.

In the process, the scientific picture of the universe came to be so far removed from the familiar that it seemed to lose touch with reality. There arose what might almost be described as a "scientific mythology."

The largeness of a galaxy is beyond realistic description. Our own Milky Way Galaxy is made up of about one hundred thirty billion stars of varying size in a complex spiral structure, within which the Earth and all it contains is less than a speck of dust. And yet astronomers reach beyond that to systems of galaxies, to billions of galaxies in colossal expansion or equally colossal renewal; to a beginning and ending that are separated by tens of billions of years; or to no beginning at all, and no ending.

But stars can at least be seen. With a telescope, galaxies can be seen. Inadequate as it is, there is at least the evidence of our eyes.

Not so in the case of the indefinitely receding horizon of the ultrasmall. The smallness of an atom is also beyond realistic description. A single atom is so small that two hundred fifty million of them, side by side, will barely span a distance of one inch. An object so small, impalpable, invisible, would seem myth enough, but scientists reached beyond.

Particles far smaller than an atom were brought into the scientific picture; particles so small that a hundred thousand of them, side by side, would but suffice to span a single atom.

And the oddest of these particles was one that had no size at all, so to speak, no handle of any sort by which it could be seized. It was so small, so entirely a "nothing" that it could penetrate endless thicknesses of dense matter seemingly unaware that the matter was there at all.

This nothing-particle was named the neutrino and the only reason scientists suggested its existence was their need to make their calculations come out even.

The suggestion of the existence of the neutrino was made with the utmost reluctance, for even scientists felt that the conjuring up of a nothing-particle was going too far. Surely this must be sheer mythology.

And yet the nothing-particle was not a nothing at all. To the astonishment of all—even of the scientists themselves—it turned out to be real. The need to make calculations come out even had conjured up something absolutely unbelievable—but real.

The manner in which a nothing-particle was first reluctantly proposed and then triumphantly displayed is one of the truly exciting adventures of science. It is this adventure which I will try to describe in this book.

To do so properly I will have to reach backward to the dawn of modern experimental science, so that we can trace the centuries-long chain of reasoning that eventually brought the concept of the neutrino to birth.

THE NEUTRINO

1

MOMENTUM

GENERALIZATIONS

We like to imagine sometimes that there are fantastic places where anything can happen. The most famous descriptions of such places are in Lewis Carroll's book *Alice's Adventures in Wonderland* and its sequel *Through the Looking-Glass*.

In these books rabbits, frogs, and caterpillars talk, and so do playing cards and chessmen. Alice grows larger and smaller without warning, flies without wings, and falls a long distance without hurting herself. She meets legendary monsters, observes nursery rhymes come true, and watches a woman turn into a sheep.

Such places are fun to visit, but surely no one would want to live there.

It is much more comfortable to live in a universe where only certain things can happen and where you have learned, from experience, what those things are.

We accept the fact, in other words, that the universe works according to certain rules and it is the business of the scientist to try to find out what those rules are.

The scientist, in his attempt to do this, observes events carefully, and if he notes that something happens time after time after time, he may come to the conclusion that it cannot happen otherwise. He has then made a rule.

If a rule is deduced, there are several ways of judging how useful it is. For instance, the more general it is—that is, the more cases it covers—the better rule

1

it is. And the few exceptions you can find, the better rule it is. Indeed, to have a really good scientific rule, you would want no exceptions at all.

You might begin, as an example, with a rule to the effect that: All green rocks thrown up into the air will come down again.

Such a rule is useful for it tells you what to expect of green rocks and what not to expect. If you throw a green rock into the air, you can feel confident that it will come down again and you can guide your actions on that basis.

Experience will also tell you however that: All blue rocks thrown up into the air will come down again. And also: All gray rocks thrown up into the air will come down again.

It isn't hard to see that the rule can be made more general if we stated it: All rocks thrown up into the air will come down again.

You might even be tempted to make the rule still more general by saying: Anything thrown up into the air will come down again. Or, since it is always best to state a rule as simply as possible: All that goes up must come down.

When you can make a rule cover a very wide set of events it becomes tempting to call it a "law of nature." The phrase is often used but is misleading because it implies that such a rule is a "law" that has been passed by some heavenly congress and that there is a divine ukase against its being broken.

It is better, I think, to call a very general rule simply a *generalization*. This term stresses the fact that the rule is man-made; it is deduced by human beings from a number of observations. A generalization can be broken very easily for it might prove wrong. It might have its exceptions. It might hold only under certain conditions.

For instance, almost everyone will agree that: "All that goes up must come down" is a very broad and useful generalization. But is it a "law of nature"?

Rocks, baseballs, golfballs, bricks, and many other

things hurled into the air will indeed come down, but what about a bird or an airplane?

To be sure a bird must come down eventually, even if only after it dies; and an airplane must come down after it has run out of fuel, if no sooner. Nevertheless a bird and an airplane don't come down in quite the same manner a rock does. Shall we change the generalization to: "All that goes up must come down, but not necessarily right away"?

Then what about a puff of smoke or a nonleaking helium-filled balloon? Both these objects float in air and needn't ever come down. Or what about a fine dust particle? Shall we change the rule to: "All that goes up must come down, but not necessarily right away, and only if it is heavier than air or if we are performing the experiment in a vacuum"?

Yet there's still the case of a rocket that shoots into space at seven miles per second. At that velocity, the rocket can assume an orbit about the Sun independently of the Earth and need never return to the Earth's surface. Shall we change the rule to: "All that goes up must come down, but not necessarily right away, and only if it is heavier than air or if we are performing the experiment in a vacuum, and even then only if its velocity is under seven miles per second"?

As you see, the generalization that started in so simple a fashion is becoming more and more cumbersome. It is not easy to find a supremely useful generalization and a scientist who finds one is assured of renown.

As an example of an extremely useful generalization, I will present one advanced in 1687 by the English scientist Isaac Newton. One way of stating this generalization is: "The acceleration produced by an unbalanced force acting on a body is proportional to the magnitude of the net force, in the same direction as the force, and inversely proportional to the mass." In mathematical symbols, this generalization is most easily expressed: $a = f/m$.

This *second law of motion* (Newton also presented a

first law of motion, and a third law of motion) can be
made to apply to all motion of any sort. As you can
well imagine, however, working out this interrelationship of acceleration, force, and mass required more
careful observation and more subtle insight than did
the "What goes up must come down" generalization.

In this book, we are going to be particularly concerned with a group of generalizations that are the
most fundamental ones we know in science. These
generalizations involve the very reverse of motion; they
involve changelessness.

BOUNCING BILLIARD BALLS

It would be extremely useful if we could find specific
properties of the universe which do not change, despite
all the general changeability we see about us. A bit of
changelessness amid all this flux would give us something to anchor to. And in fact, whether we are aware of
it or not, we rely confidently on particular events taking
place because we do assume certain properties of the
universe to be unchangeable.

The billiards expert, for instance, is reasonably certain about the outcome of his shots, provided he
strikes the ball accurately with his cue (which, since
he is an expert, is to be expected) and provided there
is no sudden earthquake or similar disruption at the
moment of striking.

What makes him so certain? How does he know
that the balls will do exactly what he expects them to
do?

Experience, of course, is the basic reason.

But what is there about the activity of moving
billiard balls which is so regular and unvarying that after
observing several hundred or several thousand shots, the
player can "get the idea"? You can, after all, play the
horses or the stock market all your life and never be
able to predict what will happen next with the same
certainty that the billiards expert can.

Apparently, moving billiard balls represent a simpler

system than do either moving horses or moving stock market prices, and a useful generalization is therefore easier to extract from the behavior of billiard balls. Let's see what that generalization might be.

Imagine a billiard ball moving along the surface of the table in the simplest possible manner, with no unusual spin of any sort, at a steady velocity of 10 centimeters per second or, to abbreviate that, 10 cm/sec.*

Suppose, further, that this billiard ball strikes a stationary billiard ball squarely dead center. A careful observer might note that the second billiard ball, stationary at the moment of impact, starts off at once, while the first billiard ball, previously moving, comes to a complete halt. In this case, it will be found that the second billiard ball is now moving at 10 cm/sec in precisely the same direction that the first billiard ball had been moving.

It might also be that after impact both billiard balls move forward. In that case, each will move at a velocity less than 10 cm/sec, but the sum of the two velocities will be 10 cm/sec.

By dint of continuing observation, it has proven the experience of mankind that in such a dead-center collision, the sum of the velocities of the balls after impact is the same as the sum of the velocities before impact. (Actually, there is a slight slowing effect as a result of the friction of the ball against the tabletop and of the resistance of the surrounding air, but these effects can be ignored for the moment.)

In short, total velocity remains unchanged, while other facts, such as the position and individual velocity of each ball does change. Total velocity seems to be "conserved."

The value of such a generalization is that it eliminates all sorts of events from the realm of the possible. You can be sure that neither ball will move more quickly

---

*The *centimeter* is a unit of the metric system commonly used by scientists. It is equal to about ⅖ of an inch so that 10 cm/sec is equal to about 4 inches per second.

than a certain limit. Furthermore, if you know the velocity of one ball in such a two-ball system, you have fixed what the velocity of the other ball must be. You have put one aspect of the universe in chains, so to speak; you have tamed it and made it understandable and predictable.

But will this "conservation of total velocity" work in all cases, or only in the case I have just outlined?

What, for instance, if the speeding ball strikes another moving ball and not a stationary one? Suppose that while the first billiard ball is moving 10 cm/sec to the north (let us say), the second is moving 10 cm/sec to the south, and the two meet head-on. We would expect billiard balls to rebound under these conditions but what if the balls were made of wax or putty so that on colliding they would flatten and stick to each other? It would turn out, then, that the two speeding balls would come to a dead halt at the point of contact and, in that case, what would have happened to all that velocity? Certainly, if velocity disappears, one can scarcely view it as being conserved.

It is the experience of careful observers that in order to talk of conservation of velocity, one must consider the direction in which an object moves and not only its speed. Suppose we say that a motion of 10 cm/sec to the north can be represented as a velocity of $+10$ cm/sec, while the same motion to the south is -10 cm/sec. In such a case, the net velocity of the two moving balls taken together is not 20 cm/sec at all. It is $10-10$ or 0 cm/sec.

Consequently, if the two balls collide, stick, and remain motionless at the point of impact, there is no change in net velocity at all. Zero remains zero.

If the balls were true billiard balls and properly elastic so that they bounced easily, the situation would be different. Each ball would suddenly reverse direction of motion. The northward moving ball would rebound southward, its velocity changing from $+10$ cm/sec to -10 cm/sec. The other ball would bounce back also and change from —10 cm/sec to +10 cm/sec. The net velocity is still zero.

It might be, if the balls were not perfectly elastic, that one ball would change velocity from +10 cm/sec to —6 cm/sec. In that case the other would change from —10 cm/sec to +6 cm/sec. There are many possibilities of different changes as a result of such a head-on collision, but these possibilities would be limited by the fact that there must nevertheless be "conservation of net velocity." The use of the word "net" makes the generalization fit balls moving in different directions.

There is still, however, room for greater complication. What if a moving billiard ball hits a stationary one, but not squarely? The collision is off center. What then?

If you have ever watched billiards, you know the answer to that one. The balls change direction. The stationary ball is set into diagonal motion (to the left, if it has been hit right of center) while the ball originally in motion changes course and moves off diagonally to the right. If one ball heads to the left of the original line of motion, the other always heads to the right; two balls, under these conditions, are simply never observed to move to the same side of the original line of motion.

Let's consider diagonal motion in two dimensions (such as the flat top of a billiard table). Such diagonal motion can always be broken up into two *component motions* at right angles to each other. This is done by drawing a line in the direction of the motion, with its length representing its speed. This line, representing the diagonal velocity, is then made the diagonal of a rectangle (see Figure 1). The size of the two component velocities can be obtained by taking the proportion of the lengths of the sides of the rectangle to the diagonal. This can be calculated if we know the angles of the geometric figure. We don't have to concern ourselves with such calculations in this book, but Figure 1 represents a simple case in which diagonal motion of 10 cm/sec

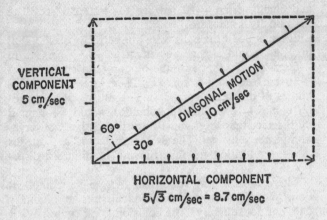

FIGURE 1. Component Velocities

has a vertical component of 5 cm/sec and a horizontal one of 8.7 cm/sec.†

Let us now go back to the off-center collision of billiard balls, in which each billiard ball takes up diagonal motion after impact. If the diagonal motion of each billiard ball is broken up into its component motions, it is found that the sum of the forward component motions after collision is equal to the forward motion before collision. In the case shown in Figure 2, the originally moving ball had a velocity of 10 cm/sec. After impact, its forward component was 2.5 cm/sec, while the forward component of the originally motionless ball was 7.5 cm/sec.

Each ball after impact has a sideways component motion as well, whereas before the collision, one ball was moving directly forward and the other was stationary. As you see in Figure 2, however, the sideways components are equal in size and opposite in direction

†A similar treatment will work in three dimensions where a particular straight-line motion can be broken up into three components at mutual right angles, each component being proportional in velocity to the lengths of the three sides of a cube of which the original motion represents the diagonal.

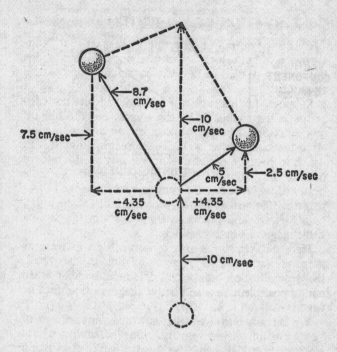

8.7 cm/sec

7.5 cm/sec

10 cm/sec

5 cm/sec

2.5 cm/sec

−4.35 cm/sec

+4.35 cm/sec

10 cm/sec

FIGURE 2. Off-Center Collision

(+4.35 cm/sec and −4.35 cm/sec in the case shown in the figure) so that the net sideways motion is zero. It is for this reason that under the conditions described, one ball must angle off in one direction after impact, and the second ball in the other direction. If both moved left of the original line of motion, for instance, a net sideways motion would have been created where none had existed before.

All this holds true no matter how many billiard balls are involved in a collision or in how many different directions they go. The net velocity in any particular direction is the same after the collision as before.

CONSERVATION OF MOMENTUM

By now, you may be starting to suspect that "conservation of net velocity" will hold under all conditions but wait— We have not yet considered all possible situations.

Suppose, for instance, the billiard ball hits the edge of the billiard table. It bounces back, but the billiard table, motionless before the collision, remains motionless after the collision as well. There seems nothing with which to balance the change in velocity of the billiard ball. If the billiard ball strikes the edge of the table head-on, its velocity of $+x$ cm/sec, will change to $-x$ cm/sec. If it strikes the edge of the table at an angle, both forward and sideways components of the motion will change sign in this fashion.

Here net velocity is not conserved and as soon as even one case of such nonconservation is discovered, the generalization breaks down. It must be abandoned and, if possible, a new and better generalization must be found.

We can ask ourselves what is wrong with our law of "conservation of net velocity." One possibility is that we have been living in a fool's paradise by considering too restricted a set of conditions. All our colliding and bouncing billiard balls have been equal in size. What if we considered balls of different size or, to use a more appropriate word, balls of different *mass?*

The word "mass" was used earlier when I gave you Newton's second law of motion and, indeed, mass is best defined by means of that second law. You might say that mass is the ratio of the force applied to a body to the acceleration produced by the force on the body.

Here on the surface of the Earth, however, and under ordinary conditions, the mass of an object is proportional to its weight. For this reason, mass is usually measured by weighing and we can be pretty safe in saying that the heavier an object, the more massive it is; the lighter an object, the less massive it is.

In the metric system, mass can be measured in *grams,* where 1 gram is equivalent to about 1/28 of an ounce. A *kilogram* is equal to 1000 grams, and is the equivalent of about 2.2 pounds. "Gram" may be abbreviated as *gm* and "kilogram" as *kg*.

Let us consider two balls then, a moving ball of 70 grams mass and a stationary one of 35 grams. If the 70-gram ball is moving at 10 cm/sec and strikes the stationary ball squarely, the latter may shoot forward at 8 cm/sec, while the previously moving ball will continue forward at a velocity of 6 cm/sec. The net velocity forward before the collision was 10 cm/sec, but after the collision it is $8 + 6$, or 14 cm/sec (see Figure 3). Where has the additional velocity come from?

It may be that velocity doesn't count as much when it is associated with a smaller mass. Since the 35-gram ball is only half as massive as the 70-gram ball, perhaps only half its velocity should be counted. Half its velocity is 4 cm/sec. That, added to the 6 cm/sec velocity of the 70-gram ball makes a net velocity of 10 cm/sec after collision as well as before.

Another way of achieving this same result is to multiply the velocity by mass where the product of mass times velocity is called *momentum*. Originally, the 70-gram ball was moving at 10 cm/sec. We may say therefore that the original momentum of this ball was 70 gm multiplied by 10 cm/sec, or 700 gm-cm/sec.‡

After the collision the 70-gram ball is moving at 6 cm/sec for a momentum of 420 gm-cm/sec, while the 35-gram ball is moving at 8 cm/sec for a momentum of 280 gm-cm/sec. The total momentum is therefore 700 gm-cm/sec after collision as well as before.

This is what turns out to be the case in all collisions. It is not velocity alone (or component velocities) that

‡Notice that in multiplying such quantities, the units must be multiplied as well as the numerals. The *gm-cm/sec* is therefore a "unit of momentum."

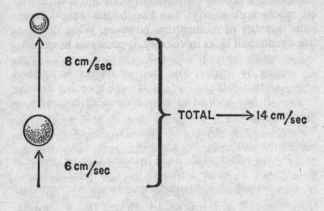

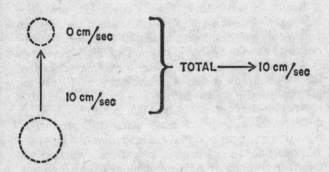

FIGURE 3. Manufacturing Velocity

must be considered before and after the collision, but the momentum (or component momenta). It is momentum, then, that is conserved, something that was first pointed out clearly by the English mathematician John Wallis in 1671.

Substitute momentum for velocity and a great deal that would otherwise be puzzling becomes easy to understand. Give a basketball a push with the toe of your foot and it will move off at a certain velocity. Give a cannonball a precisely equivalent push and it will move off much more slowly. The two bodies gain just the same quantity of momentum, however. What the massive cannonball lacks in velocity, it possesses in mass.

Or, again, suppose a loaded rifle is suspended from the ceiling by a wire, and its hair trigger is touched lightly so that it fires. The bullet goes speeding out the muzzle. It had had no motion a moment before; now it is moving very rapidly, let us say, to the right. To balance that motion, the remainder of the rifle must move to the left.

If it were velocity that were being conserved, the rifle would have to move leftward just as quickly as the bullet was moving rightward, but anyone who watches such an experiment will see this is not so. It is momentum, velocity times mass, that is being conserved. The rifle is much more massive than the bullet and need move correspondingly less quickly.

Now we have the answer to the problem of the billiard ball hitting the edge of the table. The billiard ball alters its momentum when it strikes, and the table must alter its momentum in just the opposite fashion. The billiard table is much more massive than the ball, however, and its momentum change requires a much smaller change in velocity. Indeed, the billiard table is fixed firmly to the body of the Earth itself (by friction or otherwise) so that, in effect, the whole Earth alters its momentum to balance the change in the billiard ball. The Earth is so massive that the balancing change in its motion is too small to be measurable.

In fact, we can conjure up a startling picture in connection with a bouncing ball, if momentum is to be conserved. As the ball drops downward, the Earth rises to meet it, and as the ball bounces upward, the Earth moves downward again. In short, the Earth bounces with the ball, but to as much less an extent as it is greater than the ball in mass. In view of the Earth's gigantic mass, it is no wonder that its bouncing, and its myriad other motions in balancing response to all the motions proceeding on its surface, goes undetected.

There is a similar explanation to the fact that any object moving along a long straight surface will eventually come to a rest. Its momentum will not have disappeared. Through friction, the earth will gradually have picked up that momentum.

Again, when a car starts and accelerates to high speed, its rubber tires push against the Earth which moves in the opposite direction, but, because of its huge mass, at an indetectable velocity. Anyone who has tried to start his car while it was resting on slick ice where the rubber tires can't transfer an opposite momentum change to the Earth, knows that the car won't gain momentum without such a transfer.

It is the indetectable involvement of the Earth that has made it difficult to see that momentum is conserved. Once that phenomenon is appreciated, we can feel reasonably confident in advancing the generalization commonly known as the *law of conservation of momentum*.

We can express this generalization quite simply as follows: *The net momentum of a closed system remains constant.*

By a closed system we mean any body or collection of bodies which is not influenced in any way by the universe outside. Actually, no collection of bodies is truly isolated so that the law of conservation of momentum might be considered as true, in full and complete accuracy, only for the universe taken as a whole. Systems smaller than the universe can, however, often be considered virtually isolated without serious inaccuracy. For instance, the billiard balls, together with the table, the cues and the players, can be considered an isolated system as long as there are no earthquakes during play or the family cat does not jump on the table in pursuit of a ball, and so on.

It is important to realize that the law of conservation of momentum (like all the other conservation laws I will mention in this book) is only the result of experience. It is not something deduced by logic from some basic truth, but is a conclusion based on observation.

Strictly speaking, we cannot say that momentum must be conserved under all conditions. We can only say that momentum happens to be conserved under all conditions that anyone has observed and to the utmost accuracy that anyone has been able to measure.

But how sure, then, can one be that a violation of the law cannot take place? All we have is experience to go on and our experience may not be wide enough. Earlier in the chapter, it seemed for a short time as though we had a law of conservation of velocity, but as experience widened, that fell by the wayside. May this not be true for conservation of momentum too? If not right now, then someday?

The answer is: Yes, of course. Something of this sort can always happen and in recent years some important conservation laws have indeed broken down quite unexpectedly. (I will describe one such case later in the book.)

Nevertheless when a phenomenon is observed which seems to make it appear that an important generalization may not be true, it behooves scientists to study the phenomenon very carefully. It may be that the phenomenon can be interpreted in such a way as not to constitute a violation at all. If this can be done, so much the better.

In the case of the law of conservation of momentum, for instance, so many observations of such vast variety have agreed with it, from cosmic systems of stars to infra-minute systems of subatomic particles, that scientists would find it difficult indeed to accept any violation. If they came across a phenomenon that seemed like a violation, they would be ready to accept almost any explanation of the phenomenon that would save the generalization.

This is not because the law of conservation of momentum is a "sacred cow," but only that it has proved so enormously useful over nearly three hundred years that scientists would naturally wish to retain it if at all possible.

CONSERVATION OF ANGULAR MOMENTUM

Motion need not involve a change in position. A billiard ball might spin rapidly without moving from its spot, and yet it would not be fair to consider such a billiard ball as stationary either. A ball might, for that matter, both move in a straight line and spin as well.

Any body that moves in a curve, whether only itself is involved (as when the Earth rotates about its own axis) or when an outside body is involved (as when the Earth revolves about the Sun), possesses *angular velocity*.

You would expect that a body would also possess *angular momentum* and so it does. You might also suspect that angular momentum would be angular velocity times mass, in analogy to the situation in the case of ordinary momentum,§ but here you would be wrong. Such a definition of angular momentum would make conservation of that property impossible.

Imagine, for instance, that you are standing on a small turntable, holding a heavy weight in each hand and keeping your arms close to your side. You are set to turning, and if the turntable is reasonably frictionless, you will continue turning for a considerable length of time at a reasonably constant angular velocity. (You will make, let us say, two turns per second.)

If angular momentum were the product of mass and angular velocity, and if angular momentum were conserved, then you could alter angular velocity by altering your mass. If someone were to take the weights out of your hands, for instance, the mass of material on the turntable would decrease and your angular velocity

§Sometimes "ordinary momentum" is referred to as *linear momentum* to indicate it results from motion in a straight line as opposed to angular momentum arising from motion in a curved line. Nevertheless, linear momentum is almost always referred to simply as "momentum" and that is how I shall refer to it in this book.

would increase. If additional heavy weights were tucked under your arms, your angular velocity would decrease.

If, however, mass and angular velocity were all that were involved, you would not expect to alter angular velocity at will by some simple change that did not involve mass. Yet this is exactly what you can do.

Suppose you are standing on the turntable, holding your weights close to your side and making two turns per second. Extend your arms (with the weights in the hands) as far as you can and suddenly your angular velocity slows and you move perhaps at the rate of only one turn per second. Draw your arms close to your side again and your angular velocity speeds up to its original amount.

What has happened? The total mass upon the turntable has not increased or decreased simply because you've extended your arms. Then why does angular velocity change?

It must change in response to some property alteration in the system, that does not involve quantity of mass. The logical answer is to suppose that the distance of mass from the axis of rotation is involved in angular momentum. A part of the mass on the turntable (your arms and the weights you carried) increased its distance from the axis of rotation and if this distance is involved in angular momentum, then it is to be expected that angular velocity would decrease in order to balance the distance increase. When arms and weights were drawn in again, their distance from the axis of rotation decreased again, and angular velocity increased to balance that.

If angular momentum is defined as the product of mass, angular velocity, and the square of the average distance of the mass from the axis of rotation, then angular velocity is conserved.

Under these conditions, we can say that there is a *law of conservation of angular momentum* which no one has ever seen violated. It can be stated: *The net angular momentum of a closed system remains constant.*

I say "net angular momentum" because angular velocity can be in different directions just as linear velocity can. The directions of spin, or angular velocity generally, are commonly distinguished as *clockwise* and *counterclockwise* by comparison with the familiar motions of the hands of a clock. Thus, if the Earth were viewed from high above the north pole, it would seem to be rotating counterclockwise.

If two bodies, otherwise identical, were spinning in opposite directions, one clockwise at 10 turns per second and one counterclockwise at 10 turns per second, we could consider the angular velocities to be +10 turns per second and —10 turns per second respectively. The net angular velocity of the system composed of both balls would be 0. Since the balls are of equal mass and of identical shape and constitution, the net angular momentum of the system composed of both balls is 0. The two balls can collide in such a way as to allow the spin of one to cancel the spin of the other. Neither ball might be spinning after collision and that would not violate the law of conservation of angular momentum.

Similarly, spin can be introduced into a nonspinning system, provided one part of the system is given a

COUNTERCLOCKWISE   CLOCKWISE

FIGURE 4. Counterclockwise and Clockwise Motion

clockwise spin and the other part a balancing counter-clockwise spin.

It is important to remember that, despite the similarity in names and in manifestations, the two conservation laws of momentum and of angular momentum are completely independent. You cannot convert the net forward motion of a closed system into a net clockwise spin, or vice versa—at least no one has ever observed it done.

2

ENERGY

CONSERVATION OF MASS

In considering momentum, we are really dealing with three quantities: 1) velocity, 2) mass, and 3) the product of the two, which is momentum itself.

Of these three, we have considered two from the standpoint of conservation: momentum, which is conserved, and velocity, which is not conserved. What about the third item, mass?

If we were to observe the world casually, it might well seem to us that there is clear evidence for the nonconservation of mass. Wood burns and leaves ash behind, but the ash contains much less mass (or, in other words, is less massive) than the original wood. Most of the mass of the wood would therefore seem to have disappeared. If we burned a candle completely, in fact, it would seem as though all its mass had disappeared.

On the other hand, if a piece of iron is allowed to rust completely, the rust that is formed is considerably more massive than the original iron. Here mass appears to have come into being out of nowhere.

Since mass is an inalienable property of matter, as far as we know, so that we can't have one without the other, we can equally well think of the processes of burning or rusting as demonstrating the disappearance or appearance of matter.

However, a conservation law can't be tested properly in an open system. We found that out when we tried to deal with the billiard ball rebounding from the edge of the table, without taking into account changes in momentum of the table itself.

Clearly, a burning log or candle, or rusting iron, taken by themselves, form an open system for they interact a great deal with the universe round and about. As the log or candle burns it gives off gases and vapors and these mix with the Earth's atmosphere. Certainly before any conclusions as to conservation of mass can be reached, the mass (if any) of those gases and vapors should be considered.

The process of rusting is a bit more subtle, but we might suspect that some part of the air might join the iron in the process and that the mass of that part of the air must be taken into account before we can decide whether mass is conserved or not.

Up to the eighteenth century, chemists did not truly appreciate the material nature of air and of gases generally. It was possible, therefore, to assume that gases did not have mass, or perhaps that their mass, if present, was small enough to be neglected. The eighteenth century, however, saw tremendous work done on the properties of gases and it became clearly impossible to dismiss gases so easily.

The turning point came with a French chemist, Antoine Laurent Lavoisier, who described his conclusions in a chemistry textbook he published in 1789.

Lavoisier carried on chemical reactions such as burning and rusting in sealed containers from which vapors could not escape and into which air could not enter. Mass could neither enter nor leave the system which was thus closed with respect to that property.

Lavoisier weighed the entire container, with its contents, before and after the reaction. He found no change in mass large enough to be detected by his measuring devices.

His results were confirmed by later experimenters who used more and more delicate methods of measuring mass. By the opening of the twentieth century, measurements showed that mass held constant to at least 1 part in 100,000,000.

It seems fair enough, then, to say that Lavoisier had established the *law of conservation of mass;* or, as it is sometimes called, the *law of conservation of matter*.

Mass differs from the other conserved quantities we have considered in an important respect. Both momentum and angular momentum are *vector quantities;* that is, they have direction. You can have momentum forward and momentum backward; clockwise angular momentum and counterclockwise angular momentum. This means that momentum in one part of a system can be canceled by opposite momentum in another part of the system. It means that momentum in one part of a system can be created, provided we create an opposite momentum in another part. In conserving momentum or angular momentum, we must therefore deal with net quantities obtained by subtracting the negative variety from the positive.

Mass, however, is a *scalar quantity,* one which has size but no direction. One object can be more massive than another but there is no such thing as a positive mass and a negative mass which can cancel out. To get the total mass of a system, we need only add the masses of the parts making up the system, without worrying about sign. We can talk of total mass, rather than of net mass.

We can therefore express the law of conservation of mass as follows: *The total mass of a closed system remains constant.*

CONSERVATION OF ENERGY

Velocity is involved with something other than momentum. A moving cannonball will batter down a stone wall, though the same cannonball, when stationary, will do nothing to the wall, even though it is in contact with it. The moving cannonball does *work,* while the stationary cannonball, identical in all ways save its lack of motion, does none.

To a physicist, work is defined as the motion of matter against resistance. Thus one does work in lifting a weight against the pull of gravity, or forcing a nail into wood against the resistance of friction, or breaking a stone wall and moving its fragments apart against the intermolecular forces holding all of it together.

Work, taken by itself, is clearly a quantity that is not conserved. If a weight is lifted, work has appeared out of nowhere. On the other hand, the weight will not move upward by itself. Something has to move it upward; something has to do the work. Therefore, if you want to deal with a closed system, you cannot consider work by itself; you must also include that which is performing the work.

Anything that is capable of performing work is said to be a form of *energy*. This comes from Greek words meaning "work within."

Thus, a moving mass contains energy in a way that mass alone does not, as I pointed out above in comparing the moving cannonball with the stationary one. Does this mean that the energy of a moving mass is equivalent to its momentum which, after all, is mass times velocity?

Not at all. The work done by a moving mass increases (as careful measurement shows) not as the velocity does, but as the square of the velocity. Triple the mass of a cannonball without altering its velocity and it will do three times the work. Triple its velocity, however, without altering its mass, and it will do three times three, or nine times the work.

If we represent mass by the symbol m and velocity by v, then momentum, the product of the two, is mv. The energy of motion, however, or *kinetic energy* (from a Greek word for "motion"), is most conveniently expressed as half the product of mass and the square of the velocity, $\frac{1}{2}mv^2$.*

Motion isn't the only route to work. You can break

---

*The use of exponents, as in the representation of v times v as v^2, is briefly discussed in the appendix.

a dish, for instance, not only by hitting it with a hammer, but by heating it strongly or by setting off a small explosion of gunpowder under it. You can lift a nail upward against the force of gravity by means of a magnet, or by a coil of wire carrying an electric current. In short there is a whole family of different manifestations of energy, of which some are: *electrical energy, magnetic energy, chemical energy, light energy* and *heat energy*. There is even "energy of position" or *potential energy*, which can be represented by a rock held six feet over a piece of crockery. As long as the rock is held there it does no work, but the energy is potentially there, for as soon as it is released, it drops and breaks the crockery.

Not only can energy be converted into work, but one form of energy can be converted into another. An electric current can produce magnetism and, in an incandescent bulb, it can produce heat and light, while in a motor it can produce kinetic energy. Chemical energy, which makes the burning of wood possible, is converted in the process into heat and light; and a chemical explosion can send objects flying and is thus converted into kinetic energy. Kinetic energy, through friction, is converted into heat; and if friction is used to light a match, heat is converted into light. When a storage battery is charged, electrical energy is converted into chemical energy, and when it is discharged, the opposite conversion takes place.

In this respect heat seems to be a special case. Any form of energy other than heat can, under the proper circumstances, be completely converted into heat. Heat, however, although it can be converted in part into any other form of energy, can never be completely converted. Some always remains as heat. Furthermore, if one form of nonheat energy is converted into a second form of nonheat energy, the conversion is never complete; some of the energy always ends up as heat.

Consequently it is convenient to consider energy under two subdivisions: 1) heat, and 2) all other forms, including work. It is not surprising then that heat has received special treatment and has even received a special unit of measurement of its own. (It should also be remembered that heat was studied quite carefully before it was recognized as a form of energy.)

The unit of heat is the *calorie*. It is the amount of heat required to raise 1 gram of water from a temperature of 14.5° C to 15.5° C. The *kilocalorie* is a larger unit, equal to 1000 calories.†

A more general unit of energy, used particularly for forms other than heat, is based on grams, centimeters, and seconds. If we express energy as $\frac{1}{2}mv^2$ the units of energy must be the units of mass times the square of the units of velocity. The unit of velocity we have been using is centimeter per second which we write cm/sec. The square of that unit is cm^2/sec^2. If we multiply this by the unit of the mass, the gram, we find that the unit of energy is the $gm\text{-}cm^2/sec^2$.

For greater convenience, physicists have agreed to supply a one-syllable name for the unit and for the purpose have chosen *erg* (an invented syllable taken out of the word "energy"). We can say, then, that 1 $gm\text{-}cm^2/sec^2 = 1$ erg. (The erg, by the way, is a small unit of energy. To lift a mere gram of matter the short distance of a centimeter against the pull of gravity consumes 980.7 ergs. In more common units, lifting 1 ounce the distance of 1 inch against gravity consumes 70,520 ergs.)

We can now ask an important question. When a given amount of nonheat energy of any form is completely converted into heat, is the same quantity of heat always formed? That is, will x ergs always be converted into y calories? If energy is not conserved, there is no reason why this should be so. If, however, energy is conserved, then this must be so. In fact, if x ergs always gives rise

† The Celsius, or Centigrade scale (abbreviated °C) is not the common temperature scale used in the United States for everyday purposes. We use the Fahrenheit scale (abbreviated °F). To give you a notion of the calorie in terms of ordinary units—it takes 2625 calories or 2.625 kilocalories, to heat a pint of water from 50° F to 60° F.

to y calories under a wide variety of conditions, there arises a strong presumption that energy is conserved.

The necessary experiments were carried through in the 1840s by an English physicist, James Prescott Joule. He attempted to convert energy into heat in a wide variety of ways. For example, he stirred water or mercury with paddle wheels; he compressed air; he forced water through narrow tubes; he rotated a coil of wire between the poles of a magnet; he passed electric currents through wires. In every case, he measured the energy that was used up and the heat that appeared. Even on his honeymoon, Joule couldn't resist taking time out to measure the temperature at the top and bottom of a waterfall to see how much heat had been produced by the energy of the falling water.

By 1847 he was satisfied that a given amount of nonheat energy of any sort always produced the same amount of heat. This has been confirmed any number of times since and we can now say that 41,800,000 ergs of energy is equivalent to 1 calorie of heat. (This is referred to as the *mechanical equivalent of heat*.)

In Joule's honor, 10,000,000 ergs has been set equal to 1 *joule*. We can therefore express the mechanical equivalent of heat by saying that 4.18 joules=1 calorie.

In that same decade, two German physicists, Julius Robert Mayer and Hermann Ludwig von Helmholtz, independently presented lines of argument to the effect that energy was conserved. With the backing of Joule's experimentation, this stand finally carried conviction and thus was established the *law of conservation of energy*, which is perhaps the most basic, fundamental, and important generalization science has yet produced.

Like mass, and unlike momentum, energy is a scalar quantity. You can have much energy or little energy, but you cannot have positive energy and negative energy. Energy cannot cancel out.

Thus, suppose two cannonballs of equal mass are flying in opposite directions at equal speeds. The momenta are equal and opposite so that the net momentum of the two cannonballs taken together is zero. If the cannonballs strike head-on and are not particularly elastic, they will squash together and drop to the ground, motionless.

But both cannonballs will also have possessed kinetic energy and this cannot cancel out. Yet once the cannonballs collide, they move no more; what, then, has happened to the kinetic energy? The answer is that it has been converted into another form of energy; specifically, into heat. The cannonballs may well have grown hot enough as a result of their collision to have partially melted.

We can speak of total energy, therefore, rather than net energy, and say that the law of conservation of energy can be expressed: *The total energy of a closed system is constant.*

UNIVERSAL GENERALIZATION

I want to emphasize again that the conservation laws I have been describing are not really "laws," but merely generalizations. Scientists make measurements here and there and find that whenever they measure linear momentum, angular momentum, mass or energy in a system that seems to be closed, those measurements remain constant whatever other changes the system undergoes.

The vast generalization is then made that these measurements will always remain constant under all conditions. But the words "always" and "all" are treacherous. Do we really know what will "always" happen under "all" conditions?

Even if we remain stubbornly confident about such matters here on Earth, what about the vast universe outside Earth?

Our measurements of conserved quantities are made on Earth under earthly conditions. It is bad enough to jump from those measurements to assumptions of "always" and "all" on Earth, but it is much more extreme to assume "always" and "all" for all the universe, when conditions elsewhere may be radically, and even inconceivably, different from those on the Earth.

Will energy be conserved under the conditions of the vacuum of outer space, a region far more empty of matter than anything we can reproduce on Earth? Will energy be conserved under the incredible temperatures of the interior of a star, temperatures we cannot study in detail in the laboratory?

In ancient times, philosophers had, in fact, taken it almost for granted that the "laws of nature" were not the same throughout the universe; that there was one set of "laws" for the Earth and another for the heavens.

There seemed reason for this, to be sure. On Earth, objects fell to the ground, but the heavenly bodies moved in stately circles and never fell. On Earth, objects changed, decayed, died, but in the heavens no change could be noted; the Sun was as fresh and bright each day as it had been the day before and throughout all the memory of mankind.

In modern times, however, evidence accumulated which tended to emphasize the unity of natural law. The first smashing bit came in 1687, when Newton published a book describing his three laws of motion. From them, he was able to demonstrate that the same force that pulled an apple to the ground when its hold on its branch broke also kept the Moon in its orbit about the Earth. Falling objects on Earth and circling objects in the heavens obeyed the same basic law of mutual attraction. This phenomenon received the name of *the law of universal gravitation*. The true accent in the phrase is on the word "universal."

But are the effects of gravitation truly universal? In Newton's time and for more than a century thereafter, the effect of gravitation could be studied only on the planets and satellites, so that the "law" was, in effect, despite its assumed universality, confined to the solar system.

In the 1790s, however, the German-English astronomer, William Herschel, discovered "binary stars," pairs of stars which, upon close observation, proved to be near neighbors and to be circling one another. Careful studies since Herschel's time have shown that these stars, hundreds of trillions of miles away, follow precisely, in their circlings, the law of universal gravitation as expressed by Newton.

Of course, there are vast distances of space beyond even the farthest known binary star and, indeed, beyond the farthest depth our most modern instruments can reach. Can we honestly say that the law of (supposedly) universal gravitation will hold through all the universe known and unknown?

No, we cannot. On the other hand, the evidence piling up in favor of the unity of "natural law" is impressive, and the burden of the proof has shifted to the negative. The physicists' attitude is something like this: "What we consider the 'laws of nature' may not apply in identical fashion to all the universe and at all times, but until we get good evidence to the contrary, we will assume they do."

This attitude is not based only upon the one grand fact of apparently universal gravitation. An even stronger set of observations bearing out the universality of the basic generalizations of science stems from the fact that the light from the farthest star seems to behave exactly as does the light from a gas flame a foot away.

Light displays properties which can be explained neatly by assuming it to consist of tiny waves of different lengths. In any sample of light, the presence of certain *wavelengths* and the absence of others can be used to yield information concerning the material making up the source of the light. Each chemical element, when made to glow at a high temperature, gives up a characteristic pattern of wavelengths, and by this pattern it can be identified. This process was thoroughly worked out in 1859 by a German physicist, Gustav Robert Kirchhoff. Because in this process light is spread out into a *spectrum* (that is, a band of different wavelengths arranged in order) the technique is called *spectroscopy* (see Figure 5).

FIGURE 5. Solar Spectrum

By means of spectroscopy, the light of the Sun could be made to yield data concerning its chemical makeup. Thus the Sun was found to contain the same chemical elements present on Earth. At least, various well-known chemical elements could be made to duplicate, here on Earth, the spectral characteristics of sunlight. Studies of the spectra of the stars gave evidence to show that the same elements behaving in the same way made up all the rest of the universe besides.

In 1868, when certain spectral characteristics of sunlight could not be duplicated by any known element, the English astronomer Joseph Norman Lockyer suggested the existence of a new element, not yet discovered on Earth. He named it *helium,* from the Greek word for the Sun. Eventually, though not till 1895, this Sun-element was indeed discovered on Earth.

If, then, we assume scientific generalization (and, in particular, the conservation laws) to be universal in scope, we can take a new view of astronomy. Until 1700, astronomers were confined to observing the heavens, since clearly it was impossible to reach up to the planets and stars and experiment with them. After 1700, however, astronomers were able to go beyond mere observation. For instance, they were able to deduce more and more about the structure of heavenly bodies, about their past and their future, by applying the Earth-developed conservation laws to them.

As an example, the solar system consists of bodies that spin about their axes and move about other bodies. Thus, the Moon moves about the Earth, and Ganymede moves about Jupiter, while both Earth and Jupiter move about the Sun.

If the solar system is viewed from high above Earth's north pole, it can be seen that the Earth spins on its axis in counterclockwise fashion. So does the Sun and so do all the planets with the exception of Uranus and (possibly) Venus. Furthermore all the planets without exception and all the satellites with only a handful of unimportant exceptions, revolve about the Sun or some central planet in counterclockwise fashion. This means that there is an enormous angular momentum in one direction, with only an insignificant portion of it canceled by angular momentum in the other.

This in turn means that any theory which tries to account for the formation of the solar system must explain the existence of all that angular momentum. It cannot have arisen from nothing but must have been produced somehow in the process by which the solar system formed, a process in which angular momentum in the opposite sense was unloaded on the rest of the universe.

Furthermore, if the bodies of the solar system are considered separately, it turns out that the planets, with less than 0.2 percent of the total mass of the solar system, possess 98 percent of the total supply of angular momentum. The sun, with more than 99.8 percent of the total mass of the solar system possesses only 2 percent of the angular momentum.

Any theory which tries to account for the formation of the solar system must therefore explain not only the existence of the angular momentum, but its queerly lopsided distribution as well.

It has not proved easy to satisfy the demands of conservation of angular momentum in devising theories of the formation of the solar system, but that very fact makes the existence of that conservation law infinitely useful. Without it, almost any theory of solar system formation would seem to fit the facts and there would be no way of choosing among them. With it, no theory devised so far has completely and satisfactorily explained the existence and distribution of angular momentum, but astronomers have, at least, been forced into certain fixed directions. In addition, when some theory finally arises which completely and logically explains the existence and distribution of angular momentum, there will be the strong presumption that this theory is correct, since it is not likely that two radically different theories will independently satisfy so stringent a condition as the law of conservation of angular momentum.

Here we have an example of the usefulness and power of a conservation law. We will come across a number of other examples before we are done.

THE SUN'S ENERGY

Angular momentum offers a puzzle as far as the distant past of the solar system is concerned, but at least there is no evidence that would lead us to think the angular momentum of the solar system is not being conserved from day to day at present. The law of conservation of energy, however, rested on far shakier foundations at the time it was first announced. Evidence for it was strong indeed on Earth, but the Sun seemed to be a bright and constant witness against it.

Consider the Sun. The most obvious characteristic of that body is the quantity of light and heat it delivers despite the fact that it is 93,000,000 miles from us. It lights and warms all the Earth and has done so constantly through all of history.

The energy in the form of light and heat pouring down from the noonday Sun upon a single square centimeter of the Earth's surface in a single minute is 1.97 calories. This quantity, 1.97 cal/cm^2/min, is called the *solar constant*.

A cross section of the Earth in a plane perpendicular to the radiation reaching it from the Sun is about 1,280,-000,000,000,000,000 or 1.28×10^{18} square centimeters in area.‡ Therefore the total radiation striking the Earth

---

‡Very large numbers and very small fractions, which are frequently used by scientists and will occasionally be used in

each minute is about 2,510,000,000,000,000,000 or 2.51×10^{18} calories.

Even this by no means expresses all the radiation of the Sun. The Sun radiates energy in all directions and only very little of it strikes the tiny Earth.

Imagine a huge, hollow sphere with the Sun enclosed at its center, with every part of the sphere 93,000,000 miles from the Sun. The Sun would light and heat every part of that sphere just as it does the Earth, and the area of this huge sphere would be over two billion times the cross-sectional area of the Earth. That means that the Sun radiates more than two billion times as much energy as the Earth manages to intercept.

The total energy radiated by the Sun is 5,600,000,-000,000,000,000,000,000,000, or 5.6×10^{27} cal/min. What's more the Sun has been radiating 5.6×10^{27} cal/min through all of recorded history and for an indefinitely long period before that, with only slight variations.

Here, then, is the crucial question: Where is all that energy coming from? If the law of conservation of energy applies to the Sun as well as to the Earth, then the incredibly vast supply of energy being poured into space by the Sun cannot be created out of nothing. Energy can only be changed from one form to another, and therefore the Sun's radiation must be at the expense of another form of energy. But what other form?

A person pondering the problem might think first of chemical energy, as the form of the disappearing energy. A coal fire, for example, delivers light and heat as the Sun does, when the carbon of the coal and the oxygen of the air combine to form carbon dioxide.

Can it be, then, that the Sun is nothing more than a

---

this book, are more conveniently expressed in *exponential notation*. Thus, 1,280,000,000,000,000,000 can be written 1.28×10^{18} where the small ^{18} is an *exponent*. On the chance that some of my readers are not familiar with exponential notation, I will write numbers in the ordinary fashion as well, when they are first introduced. A brief discussion of the nature and use of exponential notation will be found in the appendix.

vast coal fire and that its radiant energy is obtained at the expense of chemical energy?

This possibility can be eliminated without trouble. Chemists know quite well exactly how much energy is given off by the burning of a given quantity of coal. Suppose the Sun's enormous mass (which is 333,500 times that of the Earth) were nothing but coal and oxygen and that the two were combining at such a rate as to produce 5.6×10^{27} cal/min. The Sun would then, indeed, be a coal fire lighting and heating the solar system as it is observed to do. But how long could such a coal fire continue to burn at such a rate before nothing is left but carbon dioxide? The answer is easily determined and works out to be a trifle over fifteen hundred years.

This is a small stretch of time. It covers only a fraction of the civilized history of mankind (to say nothing of the long eons before that). Since the Sun was shining in its present fashion at the time of the height of the Roman Empire then we know without further investigation that it cannot be a coal fire, for it would be extinguished by now. Indeed, there is no known chemical reaction which would supply the Sun with the necessary energy for even a fraction of mankind's civilized existence.

Some alternatives to chemical energy must be examined, and one of them involves kinetic energy. We on Earth have a good display of the meaning of such energy every time a meteor strikes the upper atmosphere. Its kinetic energy is converted into heat by the effect of air resistance. Even a tiny meteor, the size of a pinhead, is heated to a temperature that causes it to blaze out for miles. A meteorite weighing a gram and moving at an ordinary velocity for meteorites (say, 20 miles per second) would have a kinetic energy of more than 5,000,000,000,000 or 5×10^{12} ergs—or about 120,000 calories.

A similar meteorite striking the Sun rather than the Earth would be whipped by the Sun's far stronger gravitational force to a far greater velocity. It would therefore deliver considerably more energy to the Sun. It is estimated, in fact, that a gram of matter falling into the Sun from a great distance would supply the Sun with some 44,000,000 calories of radiation. To take care of all the Sun's radiation, therefore, about 120,-000,000,000,000,000,000 or 1.2×10^{20} grams of meteoric matter would have to strike the Sun every minute. This is equivalent to over a hundred trillion tons of matter.

This works well on paper but astronomers would view such a situation with the deepest suspicion. In the first place, there is no evidence that the solar system is rich enough in meteoric material to supply the Sun with a hundred trillion tons of matter every minute over long eons of history.

Besides, this would affect the mass of the Sun. If meteoric material were collecting on the Sun at this rate then its mass would be increasing at the rate of about 1 percent in 300,000 years. This may not seem much, but it would seriously affect the Sun's gravitational pull, which depends upon its mass. If the Sun were increasing its mass even at this apparently slow rate, the Earth would be moving steadily closer to the Sun and our year would be growing steadily shorter. Each year would, in fact, be two seconds shorter than the one before, and astronomers would detect that fact at once, if it were indeed a fact. Since no such variation in the length of the year is observed, the possibility of meteorites serving as the source of the Sun's radiation must be abandoned.

Helmholtz, one of the architects of the law of conservation of energy, came up with a more reasonable alternative in 1853. Why consider meteorites falling into the Sun, when the Sun's own material might be falling? The Sun's surface is fully 432,000 miles from the Sun's center. Suppose that surface were slowly falling. The kinetic energy of that fall could be converted into radiation. Naturally, if a small piece of the Sun's surface fell a short way toward the Sun's center, very little energy would be made available. However, if all the

Sun's vast surface fell, that is, if the Sun were contracting, a great deal of energy might be made available.

Helmholtz showed that if the Sun were contracting at a rate of 0.014 centimeters per minute, that would account for its radiation. This was a very exciting suggestion, for it involved no change in the Sun's mass and therefore no change in its gravitational effects. Furthermore, the change in its diameter as a result of its contraction would be small. In all the six thousand years of man's civilized history, the Sun's diameter would have contracted by only 560 miles which, in a total diameter of 864,000 miles, can certainly be considered insignificant. The shrinkage in diameter over the two hundred fifty years from the invention of the telescope to Helmholtz's time would be only 23 miles, a quantity that would pass unnoticed by astronomers.

The problem of the Sun's radiation seemed solved, and yet a flaw—a most serious one—remained. It was not only during man's civilized history that the Sun had been radiating, but for extended stretches of time before mankind had appeared upon the Earth.

How long those extended stretches had been no one really knew in Helmholtz's time. Helmholtz, however felt this could be reasoned out. If the material of the Sun had fallen inward from a great distance, say from the distance of the Earth's orbit, enough energy could have been supplied to allow the Sun to radiate at its present rate for 18,000,000 years. This would mean that the Earth could not be more than 18,000,000 years old, however, for it could scarcely be in existence in anything like its present form when the matter of the Sun extended out to the regions through which the Earth is now passing.

It might have seemed that a lifetime of 18,000,000 years for the Earth was enough for even the most demanding theorist, but it was not. Geologists, who studied slow changes in the Earth's crust, estimated by what seemed irrefutable arguments that to achieve the present situation, the Earth must have been in existence not for merely tens of millions of years, but for hundreds of millions of years, possibly for billions of years; and that through all that time, the Sun must have been shining in much its present fashion.

Then, too, in 1859, the theory of evolution by natural selection had been advanced by the English naturalist Charles Robert Darwin. If evolution was to have proceeded as biologists were then beginning to think it must have, then, again, the Earth had to be in existence for hundreds of millions of years at least, with the Sun shining throughout that time much as it is today.

During the second half of the nineteenth century, therefore, the law of conservation of energy was shored up, with respect to the Sun, in a most controversial fashion. A plausible theory had been proposed which astronomers were willing to accept, but which geologists and biologists objected to vigorously.

Apparently there were three alternatives:

1) The law of conservation of energy did not hold everywhere in the universe and, in particular, did not hold on the Sun—in which case "all bets were off."

2) The law of conservation did hold on the Sun, and the geologists and biologists were somehow wrong in their interpretation of the evidence they had mustered, so that the Earth was only a brief few million years old.

3) The law of conservation did hold on the Sun, but there was some source of energy as yet unknown to science which, when discovered, would allow for the Sun's radiating in its present fashion for billions of years, thus reconciling physical theory with the views of geologists and biologists.§

§I have gone into the story of the Sun's radiation in such detail for several reasons. First, it demonstrates the wide and even cosmic conclusions that can be reached from a detailed consideration of a simple conservation law. Second, it demonstrates the eagerness with which science will weigh all alternatives rather than see a conservation law broken. Third, the consequences of and attitudes toward conservation laws will play a role in the history of the neutrino similar to that in the controversy over the source of the Sun's radiation. And fourth, solar radiation will, as we shall see, have a direct connection with the neutrino.

For fifty years after Helmholtz proposed his theory, no sure way was discovered for deciding among these three alternatives. When the matter was finally resolved, it was through discoveries in the realm of the ultimately small, rather than that of the ultimately large. It is to the microcosmic world, then, that we must now turn.

3

ATOMIC STRUCTURE

RADIOACTIVITY

The last decade of the nineteenth century saw a burst of physical discovery so brilliant as to bring about a virtual scientific revolution.

One of the discoveries involved came in 1896, when the French physicist Antoine Henri Becquerel discovered that compounds containing atoms of the heavy metal uranium were constantly giving off a hitherto undetected form of radiation. The radiation was detected by him through its ability to fog a photographic film. Furthermore, the radiation was strangely penetrating, for it could fog a photographic film that was covered with black paper or even with metal foil. The uranium compounds were said to be *radioactive* and the phenomenon was termed *radioactivity*.

In the decade that followed it was discovered that the radiation from uranium was of three types, which were named after the first three letters of the Greek alphabet, *alpha rays, beta rays,* and *gamma rays*.

As it turned out, the alpha rays consist of particles only about 1/60 as massive as the uranium atoms from which they emerged but still as massive as a simple atom such as that of the light gas helium. The individual *alpha* particles proved, in fact, to be closely related to the helium atom.

The beta rays are also composed of particles, but particles much less massive than atoms, only 1/1837 as massive as the least massive atom of all, that of hydrogen. *Beta particles* were found to be identical with other light particles found in an electric current forced across a vacuum. These latter particles, in view of their origin, were named *electrons*. A beta particle, then, can be considered an electron which is flying out of a radio-active atom.

The gamma rays are not particles in the strict sense of the word but are radiations that possess wavelike properties, much as ordinary light does, except that the gamma rays have wavelengths considerably shorter than those of light.

This, however, is an insufficient description of the gamma ray. It would have been sufficient at any time in the nineteenth century, but as the twentieth century opened, light waves were being given a new look.

In 1900 the German physicist Max Planck, after studying the manner in which hot objects radiated light of different wavelengths, found that he could account for the facts of radiation only by supposing that energy could not be freely radiated in any amount whatever. Instead, it had to be radiated in small packets, which he called *quanta* (singular, *quantum*).

An object might radiate one quantum of light, or two quanta, but never one and a half quanta, or two and a third. Instead of energy pouring out continuously, it comes out in chunks.

Quanta are so small, however, that under ordinary conditions they cannot be individually detected and the energy flow seems continuous. This is analogous to the situation in which a sandy beach seen from a distance will seem a continuous sheet of matter, and only a close look will uncover the actual separate grains of sand. A more extreme analogy is the case of an aluminum ingot, which appears as continuous matter under even the best microscope, but which, we know now, is made up of tiny separate atoms.

Not all quanta are the same in size. In particular, when energy is radiated in the form of waves similar to those of light, the size of the quanta depends on the wavelength. The shorter the wavelength, the larger the quanta. In ordinary light, each wave is about 1/20,000 centimeters long. This is very short on the ordinary scale but it is long enough to make the quanta of visible light very small.

The wavelengths of gamma rays are 1/5000 that of ordinary light, at most, and the gamma-ray quanta are, therefore, at least 5000 times larger than ordinary light quanta—and are quite sizable on the atomic scale.

Quanta can act, in some ways, as though they were particles, and in their role as particles, they were named *photons* (from a Greek word for "light"). Naturally, the larger the quanta the more pronounced are the particle-like properties of the radiation. Ordinary light, with small quanta, show very little of the particlelike properties, and throughout the nineteenth century light could be treated purely as wave forms with very few difficulties arising.

Gamma rays, on the other hand, with large quanta, show pronounced particlelike properties which are impossible to ignore (see page 61). In the mélange of particles that make up the subatomic world one therefore commonly includes the gamma-ray photon.

THE ATOMIC NUCLEUS

The discovery of alpha particles and beta particles forced physicists to change their fundamental notions about atoms. All through the nineteenth century, they had considered atoms to be the smallest possible particles of matter. Each different element was made up, it was supposed, of characteristic atoms that differed among themselves only in mass.

The mass of the individual atom is extremely small. It takes nearly three billion trillion of even the most massive known atoms to make up a gram of mass. Rather than deal with such numbers as one three-billion-trillionth of a gram, chemists agreed to assign the oxygen atoms an arbitrary mass of 16 and to measure the mass of all other atoms (the *atomic weight*) relative to that. The number 16 was chosen so that no atom, not even the lightest, would have an atomic weight of less than 1 on this "oxygen scale."*

By this system, the element hydrogen is made up of hydrogen atoms with an atomic weight of 1, helium is made up of helium atoms with an atomic weight of 4, sulfur of sulfur atoms with an atomic weight of 32, uranium of uranium atoms with an atomic weight of 238 and so on.†

With the discovery of radioactivity, however, it seemed that the atom, whatever its properties, could not be a smooth, featureless infra-tiny billiard ball as nineteenth-century chemists had pictured it. It had to have structure; it had to be made up of still smaller subatomic particles.

A beta particle, as I have said, has only 1/1837 the mass of even the lightest atom, while an alpha particle, although fairly massive, is much smaller in volume than an atom. Exhaustive experimentation has shown that a typical atom is about a hundred-millionth of a centimeter in diameter. This is small, indeed, but an alpha particle is so small that it would take perhaps fifty thousand of them, laid side by side, to stretch across the diameter of an atom.

The crucial step in the direction of understanding the internal structure of the atom was taken by the New Zealand-born British physicist Ernest Rutherford. He bombarded thin sheets of metal with flying alpha parti-

---

*Increasing accuracy in methods used for measuring the mass of atoms made it necessary to adopt a new scale in recent years. For one thing, it was discovered that not all oxygen atoms are alike in mass. For another, atoms of elements other than oxygen possessed masses which could be more accurately determined and which were therefore more suitable for use as a standard. In 1961, therefore, the weight of one variety of carbon atom was set at 12. This did not seriously alter the atomic weights already accepted but introduced minor changes that made for greater accuracy. The figures used in this book for the masses of atoms are all given in this "carbon scale."

†These are not the exact atomic weights, but are close approximations which will do for now. More precise values will be necessary and will be used later in the book.

cles and found that those alpha particles passed through the metal as though there were nothing there. From this he concluded that atoms were mostly empty space. Occasionally, though, an alpha particle did seem to approach something solid and veer off.

By 1908 Rutherford decided that every atom was composed of a tiny *atomic nucleus* located within the very center of the atom and taking up not more than about a trillionth of the volume of the atom. Despite the insignificant size of the atomic nucleus, however, it contained 99.95 percent of the total mass of the atom. The rest of the atom was taken up by electrons which were so low in mass that to the flying alpha particle, over 7000 times as massive as any single electron, the outer portions of the atom might just as well have been empty.‡

The outer parts of every atom are occupied by electrons; and all electrons are identical as far as we know. Electrons can, by one method or another, be knocked out of any atom, but each atom, of whatever element, yields the one known variety of electron.

All the chemical reactions studied by chemists involve the transfer, completely or in part, of one or more electrons from one atom to another. What is usually termed chemical energy might therefore better be termed "electronic energy" (but it never is).

A particular atom may end up with one or more electrons over and above its usual complement, or one or more below. In some cases, an atom may be stripped of all its electrons so that only the bare nucleus is left. A helium atom, for instance, ordinarily possesses two electrons. If those two electrons are removed, the bare

‡The atom, in this respect, bears certain resemblances to the solar system. The Sun, at the very center of the solar system, takes up as little of the volume and makes up as much of the mass of the system, as does the nucleus at the center of the atom. The regions outside the Sun, though containing numerous planets, satellites, asteroids, comets, and so on, would be to a meteor flashing through as nothing more than empty space.

helium nucleus that is left is identical with the alpha particle.

The atomic nucleus, although so much smaller than the atom, is, except in one case (that of one variety of hydrogen atom), not a single unstructured entity. With that exception, all atomic nuclei are made up of two or more subatomic particles. These are now known to come in two varieties but for the moment it will be sufficient to treat those two varieties without distinction and lump them together as *nucleons*.

On the atomic weight scale the mass of both varieties of nucleon is just a trifle over 1. We will not be very far wrong, at this stage of the story, to make the approximation of considering the *mass number* of a nucleon to be 1. If we do so, then the mass of a particular atomic nucleus is equal to the number of nucleons it contains. What's more, the mass of the nucleus may be taken as the mass of the atom to which it belongs. An electron has a mass number of only 0.00054, and we can ignore its contribution to the mass of the atom.

In the case of some elements, the nuclei of all their atoms possess a characteristic number of nucleons. All aluminum atoms occurring in nature, for instance, contain 27 nucleons in their nuclei and all have a mass number of 27, therefore. Such atoms may be referred to as aluminum-27.

The atoms of most elements, however, are to be found in two or more varieties, differing in nucleon number. Most hydrogen atoms possess nuclei made up of a single nucleon, but the nuclei of a very few are made up of two nucleons. We can speak, therefore, of hydrogen-1 and hydrogen-2. In the same way there are helium-3 and helium-4 (an alpha particle is the bare nucleus of a helium-4 atom), uranium-235 and uranium-238. Tin atoms occur naturally in ten different varieties: tin-112, tin-114, tin-115, tin-116, tin-117, tin-118, tin-119, tin-120, tin-122, and tin-124, but to possess so many varieties is quite exceptional for an element.

The different varieties of a particular element are

generally called *isotopes*. Hydrogen, helium, and uranium have two naturally occurring isotopes apiece; tin has ten; aluminum only one.

Chemists very commonly refer to elements by their chemical symbols, which usually consist of the initial letter of the name of the element, or that plus a second letter from the body of the name. Thus, hydrogen is symbolized as H, helium as He, uranium as U, and aluminum as Al. Tin is one of the few elements known to the ancients so that it is blessed with a Latin name (*stannum*) quite different from the English one, and the chemical symbol Sn is derived from that.

The mass number of a particular isotope is written as a superscript when the chemical symbol is used. Until quite recently the superscript was placed to the right, but now it is growing more common to place it on the left. Thus, hydrogen-1 and hydrogen-2 can be written H^1 and H^2, to use the older convention. Similarly we would have He^3 and He^4, U^{235} and U^{238}, Al^{27}, and, of course, Sn^{112}, Sn^{114}, and all the rest.

NUCLEAR ENERGY

This new view of the atom, which arose at the turn of the century, offered a possible new kind of answer to the problem of the source of the Sun's energy. Almost at once it turned physicists' attention to the third alternative listed on page 37.

The atoms of the element uranium (and of thorium, another heavy metal) are constantly emitting alpha particles at a terrific velocity that averages about 12,500 miles per second. The alpha particle, therefore, has a kinetic energy of about 0.000013 or $1.3 \times 10^{+5}$ ergs. Considering what a small unit an erg is in the first place, you might be tempted to shrug off a few millionths of that quantity as insignificant. However, it is a tremendous quantity of energy to emerge from a single atom.

To make this fact easier to grasp, let me introduce a new unit of energy, one that is considerably smaller than the erg.

Physicists must routinely pump energy into subatomic particles and they do so by subjecting such particles (electrons, for instance) to the action of an electric field. The driving force of an electric field, which makes the subatomic particle move more quickly and therefore adds to its kinetic energy, is measured in *volts*. (This unit is named for Alessandro Volta, an Italian physicist who constructed the first electric battery in 1800.)

An electron acted on by an electric potential of one volt gains a fixed amount of energy. This amount is called an *electron-volt* and is usually abbreviated *ev*. A thousand electron-volts, or a kilo-electron-volt, is abbreviated *Kev;* a million electron-volts, *Mev;* a billion electron-volts, *Bev.* (In Great Britain, what we call a billion electron-volts is called a giga-electron-volt and abbreviated *Gev.*)

One electron-volt is equal to 0.000000000001603 or 1.603×10^{-12} ergs. This is a little over a trillionth of an erg, and is a quantity that is well adapted to express energy changes on the atomic and subatomic level.§ Suppose, for instance, you allowed carbon to combine with oxygen to form carbon dioxide. Each gram of carbon combining in this fashion would liberate 7807 calories, but what if you wished to consider the energy produced by the combination of a single atom of carbon in this fashion? A calorie, and even the much smaller erg, would be an inconveniently large unit for the purpose. An electron-volt, however, would be just right.

A single carbon atom, combining with two oxygen atoms to form a molecule of carbon dioxide, would liberate just a little over 4 ev.

This is typical of the amount of energy liberated by single atoms in the course of chemical reactions. Com-

§You may feel annoyed at all these different units for energy: electron-volt, erg, joule, calorie. Each, however, has its own area of use and convenience. Compare the situation to the common units of length; inches, feet, yards, and miles. Would you measure the distance between two cities in inches, or the length of a room in miles, or the width of a dime in yards?

pare it, then, with the amount of energy pumped into the alpha particle given off by a uranium atom. That tiny quantity, 1.3×10^{-5} ergs, comes out to equal 8,000,000 ev, or 8 Mev.

A single atom shooting out a subatomic particle in the course of its radioactivity is emitting fully two million times as much energy as that same atom might produce in the course of a typical chemical reaction. Why should this be?

From the twentieth-century picture of the atom, a reasonable answer can be given. Ordinary chemical reactions, as I said, involve changes in the arrangements of the electrons of an atom and the arrangements of those light particles can easily be made at the cost of a few ev of energy.

Radioactive changes, however, such as the emission of alpha particles, take place as a result of alterations in the arrangement of nucleons within the nucleus. These nucleons are much more massive than electrons, and are crowded together with unimaginable tightness. The energies holding them in place are millions of times as great as those holding the electrons. When a nucleonic rearrangement liberates energy, that energy is produced in correspondingly bigger packets.

One can speak then of *nuclear reactions,* as distinct from ordinary chemical reactions, and *nuclear energy,* as distinct from ordinary chemical energy. Radioactivity is one manifestation of nuclear energy, and the first to be discovered.

Well, then, could nuclear energy—never dreamed of in Helmholtz's time—serve as the source of the Sun's steady and seemingly endless radiation? Spectroscopy (see page 29) had provided ample evidence that the Sun is, in actual fact, made up very largely of hydrogen. What could be done with that knowledge?

It took awhile for physicists to learn enough about specific nuclear reactions—how likely they were, how much energy they produced, and so on; but by 1938 an answer was obtained. In that year the German-American physicist Hans Albrecht Bethe worked out a series of nuclear reactions that were all possible under the conditions which were thought to prevail in the interior of the Sun. The net result of these reactions was that four atoms of hydrogen were converted to one atom of helium with the liberation of about 27.6 Mev.

If this were so, then for how long could the Sun radiate at its present rate, if it consisted of pure hydrogen to begin with, and if it converted hydrogen to helium in sufficient quantities to produce energy at the necessary rate? The answer is: about a hundred billion years.

If we take nuclear energy into account, then, the problem of the Sun's energy is disposed of. The Sun need not be contracting. Nor need geologists and biologists feel cramped with respect to the lifetime of the Earth.

At the moment the most generous estimates of the Earth's lifetime is about five billion years, but the Sun can easily have been radiating at its present rate for all that time without appreciably altering its appearance and without making very much of a dent in its hydrogen fuel supply. It can keep on going for tens of billions of additional years, in fact.

To add point to this, mankind soon learned to tap nuclear energy on his own and eventually to perfect the terrible hydrogen bomb, which makes use of nuclear reactions similar to those that power the Sun.

4

MASS - ENERGY

THE NONCONSERVATION OF MASS

The new view of atomic structure had thus immeasurably strengthened the belief that the conservation laws applied not only to the ordinary world about us, but also to the immensely large world of the astronomer. Having accomplished this, however, the atom presented the physicist with a new horizon in the opposite direction. Did the conservation laws of everyday life apply also to the unimaginably tiny world of the atom? Did the same basic generalizations apply for the smallest aggregates of matter as for the largest?

In many ways, it seemed that they did.

For instance, a speeding alpha particle can be made discernible after a fashion, by having it hurtle through a *cloud chamber;* that is, through a chamber containing gas that is supersaturated with water vapor. Gas that is supersaturated contains more water vapor than it can really hold under ordinary conditions, and the vapor has a tendency to settle out in the form of liquid droplets. Such droplets form most easily about tiny solid particles which have an attraction for water molecules or which have shapes such that water molecules will fit well upon them. These are *condensation nuclei,* and ordinary air contains dust particles, bits of salt from dried ocean spray, and other materials which can act in this way. In the absence of such nuclei, vapor will not settle out as droplets unless the supersaturation is very great or the temperature is unusually low.

49

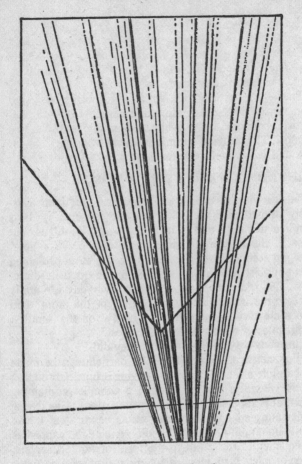

FIGURE 6. Cloud Chamber

The gas in the cloud chamber is deliberately cleaned of all dust particles so that the watervapor, although in a slightly supersaturated concentration, will not settle out as droplets. The alpha particle, as it speeds through the chamber, strikes molecule after molecule of gas, knocking electrons out of the atoms composing them.

Such atoms, now deficient in electrons, are called *ions* and these ions can serve as condensation nuclei for the formation of water droplets, where ordinary atoms could not. Thus the track of the speeding alpha particle is visibly outlined as a trail of water droplets forming about the ions it has produced.

The alpha particle may strike the nucleus of some atom, while it is still in the cloud chamber. The alpha particle bounces away, while the atomic nucleus rebounds. The moving atomic nucleus will itself form ions and will therefore leave a track of water droplets.

Physicists know the mass of the alpha particle and of the nucleus it strikes. From the curvature of the tracks in a magnetic field they can work out the velocity of the tiny objects before and after collision, and, therefore, their momenta. In all the numberless cases of colliding alpha particles and nuclei that have been observed (as well as other similar subatomic events) momentum has appeared to be conserved.

Atomic nuclei may also have spins, and thus possess angular momentum. This, too, is conserved in the course of various nuclear collisions and reactions.

It was comforting to find that a radically new kind of environment leaves untouched and unaffected the broad generalizations worked out under more prosaic conditions. Will that comfort be retained, however, when conservation of mass is taken up?

Consider, for instance, the manner in which a uranium atom emits an alpha particle. The more common variety of uranium atom, U^{238}, contains 238 nucleons and therefore has a mass number of 238. The alpha particle is the nucleus of He^4 and has a mass number of 4. When the U^{238} atom emits an alpha particle, four of its nucleons are gone and it is no longer U^{238}. It becomes an isotope of thorium containing 234 nucleons (Th^{234}). You can write the event in the form of an equation as follows:

$$U^{238} \longrightarrow Th^{234} + He^4$$

The less common uranium isotope, U^{235}, also gives up an alpha particle, forming Th^{231}. The thorium isotope, Th^{232}, (the only isotope of that element to occur in nature in significant quantities) does so too, becoming an isotope of radium (Ra^{228}). The reactions may be written thus:

$$U^{235} \longrightarrow Th^{231} + He^4$$
$$Th^{232} \longrightarrow Ra^{228} + He^4$$

In all three cases the sum of the mass numbers of the two particles produced is equal to the mass number of the original particle: $234+4=238$, $231+4=235$, and $228+4=232$.

Consider, too, the reaction that powers the Sun, which is, essentially:

$$4\ H^1 \longrightarrow He^4$$

The mass number of hydrogen-1 is 1 and the mass number of four of them is 4, which is also that of the helium-4 isotope.

Reasoning in this fashion, it might be shown that mass is conserved in all radioactive transformations and in all the nuclear reactions, generally, that involve the ordinary atoms about us.

Yet it is wrong to reason in this fashion. Earlier (see page 44) I said the mass of the nucleons was not equal exactly to 1. If we are to check the conservation of mass, it is not enough to deal in approximations. Instead, we must work with the best values physicists have managed to obtain.

For instance, the mass of the hydrogen-1 nucleus is, according to the best measurements now available, not 1, but 1.00797. That means that the mass of four hydrogen nuclei is 4.03188. The mass of the helium-4 nucleus is, however, 4.00280. When four hydrogen nuclei become one helium nucleus, a mass of 4.03188 becomes a mass of 4.00280. A mass of 0.02908 has disappeared.

The amount of this disappearing mass may seem small (it is equivalent to only about 1/34 of a nucleon) but it is far too large to be ignored. If the conservation of mass holds at all, it must hold to within the limits of measurement.

The truth of the matter is that once nuclear reactions came to be studied carefully, it was found that there was always a slight mass discrepancy between the atoms one began with and those one ended with. The law of conservation of mass, established by Lavoisier over a century earlier, did not hold after all, at least not in the atomic world. It was not a completely valid generalization.

RELATIVITY

The ability to measure the mass of individual atomic nuclei with sufficient accuracy to reveal the failure of the law of conservation of mass came about through the use of an instrument called the *mass spectrograph.* This was devised by the English physicist Francis William Aston in 1919 and reached its heyday in the 1920s.

By that time, however, the failure of Lavoisier's great generalization, which had served as the backbone of chemistry for so long, was not at all upsetting. As long before as 1905, the German-born physicist Albert Einstein (then working in Switzerland) had predicted such a failure for convincing theoretical reasons.

Einstein's theory, called the *special theory of relativity,* arose out of the inability of physicists to measure changes in the velocity of light under conditions where Newton's laws of motion predicted such changes ought to exist. Einstein undertook, therefore, to work out a broad system of generalization under which such changes in the velocity of light did not take place.

Einstein's assumptions were fundamentally quite different from Newton's, but under ordinary circumstances the conclusions deduced were virtually the same in both cases. (This is necessary, for the universe remains the universe and its properties will not change to suit a change in theory.) It was only at extremely great velocities, approaching that of light, that significant differences showed up between Einstein's and Newton's view of the universe.

The necessary extreme conditions were studied and in every case the situation uncovered agreed with Einstein rather than with Newton. Einstein's special theory of relativity is now firmly accepted by physicists and in half a century of investigation, nothing has turned up to shake it.*

A basic aspect of Einstein's theory is that no velocity can be measured which is greater than the velocity of light in a vacuum. This maximum measurable velocity is 186,282 miles per second, or 299,792.5 kilometers per second. (A kilometer is equal to 1000 meters or to 100,000 centimeters.) We won't be far wrong if we express this velocity as 30,000,000,000, or 3×10^{10} cm/sec.

Another aspect of the theory is that mass and energy could be looked upon as different aspects of the same thing. Mass behaved as though it were an extremely compact form of energy, while energy could be viewed as an extremely diffuse form of mass.

Einstein worked out the relationship between the two entities in an equation which has now become famous:

$$e = mc^2$$

where e stands for energy, m for mass, and c (the initial of celeritas, a Latin word meaning "velocity") for the velocity of light in a vacuum.

If mass is expressed in grams in this equation, and the velocity of light in centimeters per second, then energy is worked out in ergs. Since the velocity of light

---

*Newton's view of the universe, nevertheless, is much simpler to express mathematically and, under conditions of ordinary velocities, does not differ significantly from the more nearly correct Einsteinian view. The Newtonian view is therefore still used in many aspects of physics and will continue to be used —as a matter of convenience rather than correctness.

is extremely great and its square is far greater still, a tiny quantity of mass can be viewed as equivalent to a huge amount of energy. Since c is equal to 3×10^{10} cm/sec, c^2 must equal 900,000,000,000,000,000,000 or 9×10^{20} cm²/sec². If we are dealing with a mass of 1 gram, we see that, multiplying it by c^2, we have something which is equivalent to 9×10^{20} gm-cm²/sec² or 9×10^{20} ergs.

Small as the erg is, 9×10^{20} of them is a tremendous quantity. The energy equivalent of 1 gram of mass (and remember that a gram, in ordinary units, is only 1/28 of an ounce) would keep a hundred-watt light bulb burning for thirty-five thousand years.

From this view of the equivalence of mass and energy, as propounded by Einstein, it seemed necessary to assume that whenever a system gave off energy to the outside universe, it had to lose an amount of mass equivalent to that energy. If, on the other hand, it absorbed energy from the outside universe, it gained the equivalent amount of mass.

We can see now why mass seems to be conserved in ordinary chemical reactions. The energy changes involved are of such a size as to represent immeasurably small quantities of mass.

Suppose, for instance, a gallon of gasoline is burned (and remember that this is a chemical reaction that brings about a particularly large energy change). This gallon of gasoline has a mass of 2800 grams and emits, while burning, 32,000,000 calories. This is equivalent to 1,350,000,000,000,000 or 1.35×10^{15} ergs. This quantity of energy is tremendous but it is the equivalent of only 0.0000015 or 1.5×10^{-6} grams. To detect a little over a millionth of a gram in a total mass of nearly three thousand grams (1 part in 3,000,000,-000) was completely beyond the power of nineteenth-century chemistry.

The best measurements of the nineteenth-century chemists could detect no discrepancies, therefore, in the law of conservation of mass. For that matter, mass conservation can still be used as the backbone of chemistry provided only that it is applied exclusively to chemical reactions.

CONSERVATION OF MASS-ENERGY

When nuclear reactions are in question, the energy changes for a given mass of reacting material are so much greater than is true for chemical reactions, that mass-energy equivalence cannot be ignored. If mass alone is measured, then the conservation law is clearly seen to break down.

To see that, suppose we bring the mass-energy equivalence down to atomic scale. In the equation $e=mc^2$, let's deal not with 1 gram of mass but with a mass of 1 on the atomic weight scale, a mass which is approximately equal to the nucleus of the hydrogen-1 atom, the least massive atomic nucleus known.

The mass of 1 on the atomic weight scale is a small mass indeed and is equal to 0.00000000000000000000-0000167, or 1.67×10^{-24} grams. Despite the vast size of the expression c^2 (which is 9×10^{20} cm^2/sec^2), we find that the energy to which such a tiny mass is equivalent is only 0.0015 ergs.

On the ordinary everyday scale 0.0015 ergs is a petty quantity indeed, but it is equal to nearly a billion electron-volts (see page 46) and on the atomic scale that is a large amount of energy. On the basis of the best recent measurements, one can say that a mass of 1 on the atomic weight scale is equal to 0.931478 Bev. Since 1 Bev equals 1000 Mev, one can also express the mass of 1 on the atomic weight scale as equal to 931.478 Mev.

If we consider, therefore, that the mass of the hydrogen nucleus is equal to 1.00797, we can speak of this as equivalent to an energy of 0.938905 Bev, and the mass of four such hydrogen nuclei as equivalent to an energy of 3.75562 Bev. Again, if the mass of the helium nucleus is 4.00280 on the atomic weight scale, its mass is equivalent to an energy of 3.72803 Bev.

When four hydrogen nuclei are converted into one

helium nucleus, the loss of mass is therefore equivalent to 3.75562—3.72803 or 0.02759 Bev, which is in turn equal to 27.59 Mev. This energy is easily measurable and, indeed, when measured, the quantity produced by hydrogen fusion to helium is very close indeed to the theoretical prediction.

In this nuclear reaction, and in all nuclear reactions of this sort, measurements have shown that the appearance of energy is exactly counterbalanced by the disappearance of mass, in accordance with Einstein's equation.

Consequently it became customary to speak of *the law of conservation of mass-energy,* rather than the law of conservation of mass alone, or the law of conservation of energy alone. And yet the addition of the extra word seems unnecessary. One need speak only of the law of conservation of energy, provided one simply remembers that mass is a form of energy. This, in fact, is what I will do for the remainder of the book.

Let's return now to the source of the Sun's radiation. If, indeed, it is produced by the conversion of hydrogen nuclei to helium nuclei, then the vast energies produced and poured out into surrounding space must be balanced by an equivalent disappearance of mass. In that case, it may be that such disappearance could create difficulties similar to those created by the superabundant appearance of mass in the meteor theory (see page 35).

The total radiation of energy from the Sun, I have already stated, is equal to 5.6×10^{27} cal/min. This is equivalent to 3,800,000,000,000,000,000,000,000,000,-000,000, or 3.8×10^{33} ergs/sec. Dividing this by c^2, we find that this energy is equivalent to a loss of 4,200,-000,000,000, or 4.2×10^{12} grams per second. To use more familiar units, the Sun must be losing 4,600,000 tons of mass each second, or 276,000,000 tons each minute.

Does this create a difficulty? The meteor theory of solar radiation required 120,000,000,000,000,000,000, or 1.2×10^{20} grams of meteoric matter to strike the Sun each minute and this steady addition to the Sun's mass would decrease the length of the year by two seconds each year. The loss of mass in the conversion of hydrogen to helium takes place at a rate only about a thirty-millionth of the gain in mass required by the meteor theory. As a result of the Sun's loss of mass through nuclear reactions, the length of the year would increase only one second in fifteen million years, and this might well be compensated for, in part, by the gain of some meteoric matter by the Sun.

In short, the Sun's loss of mass creates no problem. The change in the length of the year is undetectable and has no practical meaning for us.

PHOTONS

Let's move into reverse now. Having considered mass in terms of energy, let's consider energy in terms of mass. A photon of light, for instance, possesses a certain amount of energy and this must, in turn, be equivalent to a certain amount of mass.

According to Planck's quantum theory, one can easily determine the energy of a photon of light from the wavelength of that light. In order to express that energy in terms of electron-volts, one must divide the quantity 0.000124 or 1.24×10^{-4} (obtained by a chain of mathematical reasoning I need not go into) by the wavelength of that light in centimeters. The longest wavelengths of visible light (deep red in color) are roughly 0.00007 or 7×10^{-5} centimeters in length, while the shortest (deep violet in color) are about half that, 0.000035 or 3.5×10^{-5} centimeters (see Figure 7).

If 1.24×10^{-4} is divided by 7×10^{-5}, we get a quotient of nearly 1.8. We can conclude then that the photon of the longest wavelengths of visible light has an energy of 1.8 ev. As the wavelength of light decreases, the energy of the associated photon increases in proportion. The shortest wavelengths of visible light, having half the wavelength of the longest, have photons twice as energetic—3.6 ev.

Since chemical reactions liberate up to about 4 ev of

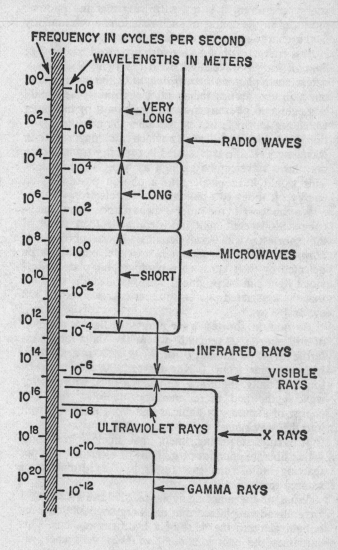

FIGURE 7. The Electromagnetic Spectrum

energy per atom, it is not surprising that the photons produced in the course of such reactions are frequently in the energy-range of visible light.

Less energetic photons are also produced, photons of light of longer wavelength than red light. We cannot detect such photons of *infrared radiation* by eye, but our skin can absorb them and feel them as heat. Still less energetic photons are associated with *microwaves* which are longer in wavelength than the infrared radiation, and with *radio waves* which are longer still in wavelength. Radio waves used in ordinary transmission may have wavelengths as high as 55,000 centimeters. This would be equivalent to a photon possessing an energy of about two-billionths of an electron-volt.

We can work in the other direction, too. Some chemical reactions emit light more energetic than those of the shortest visible wavelengths. These photons of *ultraviolet radiation,* although invisible, can easily be detected by their effect on a photographic plate. Ultraviolet light can be produced with wavelengths so short that the associated photons possess energies up to 1000 ev, or 1 Kev.

Beyond the shortest-wave ultraviolet light is a region of still shorter wavelength, where the radiation is referred to as *X radiation,* or *X rays.* X-ray photons, with energies from 1 Kev up to 100 Kev, can be found. Finally, there are the gamma rays, with shorter wavelengths and more energetic photons still. The energy of gamma-ray photons can work their way well into the Mev range.

It is not surprising, then, that nuclear reactions, which liberate energies of millions of electron-volts per reacting atomic nucleus, result in the formation of gamma rays.

What of the mass equivalence of those photons? I have already explained that an energy of 938.905 Mev is equivalent to the mass of a hydrogen nucleus. This surpasses the energy content of even very energetic gamma-ray photons.

An electron is an easier mark to shoot at. The electron

is 1/1836.11 the mass of a hydrogen nucleus and is therefore equivalent to 938.905 divided by 1836.11, or about 0.51 Mev (which can also be expressed as 510,000 ev).

A visible light photon, with the typical energy of 2.5 ev, would have a mass-equivalence equal to about 1/200,000 of an electron. Such a tiny mass is quite negligible and even on an atomic scale no great error is produced by considering the photons of visible light to be massless.

However, as one progresses down the electromagnetic spectrum toward shorter and shorter wavelengths, the photons become more and more energetic and equivalent to more and more mass. A gamma ray with a wavelength of 0.00000000024 or 2.4×10^{-10} centimeters is made up of photons possessing a mass just equivalent to that of an electron. The same devices that detect the particlelike behavior of an electron ought, therefore, to be able to detect the particlelike behavior of gamma-ray photons also.

This was demonstrated in 1923 by the American physicist Arthur Holly Compton. He found that an X-ray photon, with a mass equivalent to rather less than that of an electron, could strike an electron and make it rebound. The electron had gained energy and the photon had lost energy, precisely as though two colliding electrons had been involved. More than that, the photon acted as though it were a particle, carrying momentum, and the law of conservation of momentum was observed in the interaction of the photon and the electron.

Under the circumstances, it seemed clear that light and its related radiations had to be viewed as possessing the properties of particles as well as of waves. It was Compton, then, who suggested the name "photon" for a light quantum, making use of the "-on" suffix which had become the hallmark of the names given to subatomic particles since the discovery of the electron a quarter century earlier.

The particlelike properties of gamma-ray photons are

even more marked than are those of X-ray photons. When gamma rays are emitted in the course of nuclear reaction, their momentum must be taken into account. What's more, photons may be regarded as possessing spin and, therefore, angular momentum. In applying the laws of conservation of momentum and of angular momentum to nuclear reactions, the momentum and angular momentum of photons must be included in the calculation.

Although a gamma ray and an electron may be equivalent in mass, there is nevertheless a difference between them as far as that property is concerned. This is not a paradox, for equivalence is not necessarily identity. (A check for ten dollars may be the equivalent of ten dollars in cash, but it is not identical with it.)

Consider the mass of an electron, for instance. An electron may move at any velocity, relative to an observer, from 0 cm/sec (when it is at rest) to 3×10^{10} cm/sec (the velocity of light in a vacuum). The mass of an electron, or of any material object for that matter, varies with its velocity in such a way as to be a minimum at rest and to approach the infinite at the velocity of light.†

The mass of an object at rest relative to an observer is its *rest-mass,* and it is the rest-mass which is usually referred to when one speaks simply of "mass." When the mass of the electron is given as 9.1091×10^{-28} grams, for instance, it is well understood that that refers to its rest-mass. Electrons are often encountered which travel at velocities equal to or more than 0.99 times that of light in a vacuum and their masses are then seven or more times as high as their rest-mass.

A photon moving through a vacuum, however, travels always at the velocity of light, 3×10^{10} cm/sec, relative

---

†Einstein's Special Theory of Relativity predicts this fact and it has been amply verified by experiment. The increase in mass is quite negligible until speeds of thousands of miles per second are reached. In ordinary life we are quite safe in considering mass to be constant.

to all observers.‡ This is the cardinal point of Einstein's Special Theory of Relativity. Since a photon can never be at rest relative to any observer, one cannot measure its rest-mass directly. However, physicists find it convenient to consider the rest-mass of photons to be zero.

Despite the fact, then, that a photon can be considered as possessing the equivalence of mass, it is generally spoken of as a *massless particle,* the adjective referring to its rest-mass of zero.

The photon is not the only massless particle, as we shall see, for the core of this book deals with massless particles that are not photons. We can make the generalization that all massless particles, whether photons or not, travel at the velocity of light from the moment they are formed to the moment they are absorbed.

‡In a transparent medium other than a vacuum, photons travel at lesser velocities. Even air slows them down very slightly. When photons leave a transparent medium, however, and enter vacuum again, their velocity accelerates at once to 3×10^{10} cm/sec once more.

5

ELECTRIC CHARGE

CONSERVATION OF ELECTRIC CHARGE

For the world of the atom (as far as we have gone, at least) there are three great conservation laws that hold as tightly and as satisfactorily as they do in the everyday world we live in and in the vast astronomic universe that surrounds us.

They are:

1) conservation of momentum
2) conservation of angular momentum
3) conservation of energy

These three conservation laws all involve mass and velocity, which are thoroughly familiar quantities to us. The atom and the particles making it up brought into prominence, however, a fourth conservation law involving phenomena not quite as familiar.

It has been known since at least 600 B.C., when the Greek Philosopher Thales of Miletus investigated the phenomenon, that the fossil resin we call amber could, when rubbed, gain the property of attracting light objects. Nowadays we say that amber, when rubbed, gains an *electric charge* (or is "electrified," or gains "electricity"), the word "electric" coming from the Greek word for amber.

In 1733 the French physicist Charles François Du Fay demonstrated that there were two different kinds of electric charge, one of which was to be found in rubbed amber and one in rubbed glass.

The difference in the two types of charge can be demonstrated as follows:

Suppose two small bits of cork are suspended near each other by silk threads. Suppose further that each is touched by a piece of amber carrying an electric charge. Some of the electric charge flows into each of the bits of cork and now they are observed to repel each other. The silk threads from which they are suspended no longer hang vertically, but hang at angles that keep the corks farther apart than they were before they received the charge.

The same would happen if the two bits of cork were each touched by a piece of glass carrying an electric charge.

If, however, one piece of cork were touched by electrified amber and the other by electrified glass, the two bits would then attract each other. The silk threads would hang at an angle that would allow the bits of cork to be suspended closer together than they were before they received the charge.

It was this difference in behavior that led Du Fay to suggest the existence of two kinds of electric charge; and from it arose the generalization: *Like electric charges repel; unlike electric charges attract.*

In the 1740s the American man-of-all-abilities, Benjamin Franklin, began experimenting with electricity. He noted that if an object carrying one kind of electric charge touched an object carrying an equal quantity of the other kind, the two charges neutralized each other, leaving both objects electrically uncharged. It was as though fluid were pouring from a place where it was present in excess to a place where there was a deficit, the end result being an equal level in both, a level neither in excess nor in deficit.

Franklin suggested that the body containing the electric fluid in excess be considered as containing a *positive electric charge* and the body suffering a deficit be considered as containing a *negative electric charge*. He had no way of telling which body contained the excess and which the deficit, so he chose arbitrarily to consider rubbed glass as carrying a positive electric charge and rubbed amber a negative electric charge. This convention has been adhered to ever since.

As succeeding generations of physicists studied the behavior of electrically charged bodies, a generalization came to be accepted: *The net electric charge of a closed system is constant.*

One does not actually create an electric charge when amber is rubbed. If amber is rubbed by hand, the negative electric charge gained by the amber is balanced exactly by the positive electric charge gained by the hand. The two together still add up to zero. If the electric charge on the hand spreads into the ground and over the Earth generally, it seems to disappear, leaving the charge on the amber to appear "created," but that is an illusion.

The case is the same as with positive and negative momentum; or clockwise and counterclockwise angular momentum. We can therefore accept a fourth conservation law:

4) conservation of electric charge.

NUCLEAR REACTIONS AND ELECTRIC CHARGE

When the internal structure of the atom began to be worked out in the 1890s, physicists discovered at once that at least some of the components of the atom carried an electric charge. The electrons, for instance, which filled the outer regions of the atom, carried a negative electric charge, while the nucleus at the center of the atom carried a positive electrical charge.

The question in each case is, of course, how much? To answer that, let us consider some units of charge.

In the everyday world about us, the common unit of electric charge is the *coulomb* (named for the French physicist Charles Augustin de Coulomb who in 1785 studied the manner in which the size of an electric charge could be determined from the measured strength of its attraction or repulsion for other charges). In a

60-watt lamp, 1 coulomb of electric charge is passing through any given point in the filament every two seconds.

A much smaller unit of electric charge is the *electrostatic* unit, usually abbreviated as *esu*. The coulomb is equal to 3,000,000,000 or 3×10^9 esu.

Even the esu, however, is inconveniently large for the purpose of measuring the charge on a single electron. This charge was first measured, with reasonable accuracy, in 1911 by the American physicist Robert Andrews Millikan, and it turned out to be equal to about half a billionth of an esu. The best measurement now available is 0.000000000480298 or 4.80298×10^{-10} esu.

Rather than try to handle such an inconvenient fraction, it is most sensible to consider the electronic charge to be equal to -1, the minus sign signifying it to be a negative charge. It is particularly useful to do this because every electron, whether obtained from an electric current or from an atom of any element whatever, has a charge of exactly -1, with no difference that our most delicate instruments can detect.

Furthermore, the simplest atomic nucleus, that of hydrogen, has an electric charge of $+1$. As nearly as our most sensitive instruments can tell, the positive charge of the hydrogen nucleus is exactly as large as the negative charge of the electron (though opposite in sign, of course). Atomic nuclei larger than those of hydrogen have larger positive charges, but charges that are always exact integers. No trace has ever been found of any fractional charge, either positive or negative; at least, not so far.

The atoms of each element have a characteristic nuclear charge, different from that of the atoms of any other element. For instance, all atoms of hydrogen have a nuclear charge of $+1$, all atoms of helium have a nuclear charge of $+2$, all atoms of carbon have a nuclear charge of $+6$, all atoms of uranium have a nuclear charge of $+92$ and so on. This nuclear charge is called the *atomic number*.

Isotopes may differ among themselves in mass number but are nevertheless atoms of the same element if they are identical in atomic number. There are atoms with a nuclear charge of +1 and a mass number of 1, as well as atoms with a nuclear charge of +1 and a mass number of 2. Both types are atoms of hydrogen and can be written as hydrogen-1 and hydrogen-2, or, using symbols, as $_1H^1$ and $_1H^2$, where the superscript to the right is the mass number and the subscript to the left is the atomic number. In the same way, the two uranium isotopes can be written $_{92}U^{238}$ and $_{92}U^{235}$.

I am particularly concerned now with the conservation of electric charge so I will emphasize that quantity by writing a uranium atom of either isotope as U^{+92}. That will allow us to concentrate on that which (for the moment, at least) is most important to us.

Both uranium isotopes are radioactive. Each breaks down by eliminating an alpha particle and each is converted in the process to an atom of thorium (symbol, Th). Thorium has an atomic number of 90 and the alpha particle, which is the nucleus of a helium atom, has an atomic number of 2. We can write, then, that

$$U^{+92} \longrightarrow Th^{+90} + He^{+2}$$

We see that the original atomic nucleus has a charge of +92, and that the two resulting nuclei have a charge of +90 and +2, for a total of +92. This is a specific case of a general situation. An atom with an atomic number of x, once it emits an alpha particle, is always converted to another atom with an atomic number of $x-2$. No exception has ever been observed.

In the case of alpha-particle emission, then, we can say that the conservation of electric charge holds.

Let's see, next, how the law of conservation of electric charge would apply to the emission of a beta particle by an atomic nucleus. The beta particle is an electron, which can be symbolized as e. Since the electron has a charge of −1, we can write it as e^{-1}.

Suppose, then, that we consider the behavior of the

thorium isotopes produced by the breakdown of uranium. These thorium isotopes are not found in significant amounts in nature because they break down in their turn, and do so quite rapidly. In each case, they break down through the emission of beta particles and form an isotope of the element protactinium. Protactinium has an atomic number of 91 and its symbol is Pa. Concentrating, then, on electric charge, we can write:

$$Th^{+90} \longrightarrow Pa^{+91} + e^{-1}$$

We begin with a nucleus with a charge of +90 and conclude with two objects with charges of +91 and −1 respectively and, of course, if we add +91 and −1 we end with a sum of +90. Again, this is an example of a general rule. An atom with an atomic number of x, once it emits a beta particle, is always converted to another atom with an atomic number of $x+1$. No exception has ever been observed.

The law of conservation of electric charge holds for beta-particle emission.

Radioactive atoms may also emit gamma rays. The gamma-ray photon carries no charge at all, however. Every atom that emits a gamma ray, without known exception, retains its atomic number unchanged in the process.

In short, the law of conservation of electric charge has been found to hold for every atomic event that has ever been observed.

NUCLEAR STRUCTURE

Though beta-particle emission raised no problems as far as conservation of electric charge was concerned, there were other questions in the minds of physicists. How was it, for instance, that a negatively charged particle was emitted by a positively charged nucleus? To answer this, some conclusions had to be drawn concerning the structure of the nucleus.

The mere fact that an atomic nucleus could give off alpha particles and beta particles at all was strong evidence that the nucleus had an internal structure, that it was made up of smaller components. At least one component had to carry a positive electric charge and for a decade or so after the discovery of the electron, physicists were on the watch for some positively charged particle that was analogous to the negatively charged electron.

No such nuclear component was discovered, however. The smallest positively charged particle that could be found was the nucleus of hydrogen-1. It had a charge of $+1$ and could be represented as $_1H^1$. The electric charge, at least, was a minimum, exactly equal to that of the electron, but opposite in sign. The mass, however, was (by the best present figures) 1836.11 times that of the electron.

By 1914 Rutherford was ready to accept the hydrogen nucleus as the positively charged particle of minimum mass, present as a component of all atomic nuclei. Why this positively charged particle should be so much more massive than the negatively charged electron when both had an electric charge of exactly equal size (though of opposite signs) he could not tell. Neither could anyone else, then or now. It remains one of the unsolved problems of nuclear physics to this day.

In 1920 Rutherford suggested that this positively charged particle be called the *proton* from a Greek word meaning "first" because its large mass made it seem the first (that is, the most important) of the subatomic particles. The mass of the proton, on the atomic weight scale, is 1.00797, and no great error is introduced under most conditions if that mass is taken simply as 1.

The nucleus of hydrogen-1 was clearly made up of a single proton. All other nuclei had to contain two or more protons, but it was quickly realized that an atomic nucleus (other than that of hydrogen-1) could not be made up of protons only. Since a proton had an electric charge of $+1$ and a mass number of just about 1, all nuclei would have to have an atomic number equal to their mass number if they included only protons. But this is true only of hydrogen-1. All other nuclei have mass numbers which are greater than the atomic numbers.

Consider a nitrogen nucleus with a mass number of 14, for instance. If we were made up of protons only, it would have to have an electric charge of $+14$ and therefore an atomic number of 14. In actual fact, the electric charge is but $+7$, so that the nucleus can be symbolized as $_7N^{14}$. What has happened to the remaining seven units of charge?

Physicists thought at first the answer might lie in the presence of electrons in the nucleus. If the nitrogen nucleus contained 14 protons and 7 electrons, the 7 electrons would add very little to the mass (little enough to be ignored) but would cancel out seven of the fourteen positive charges. As a side effect, the presence of nuclear electrons would also account for the ability of nuclei to eject electrons in the form of beta particles.

This model of nuclear structure foundered on the question of *particle spin*. It was found that charged particles often gave rise to a magnetic field. This was known to happen only when the charged particles were in motion, so it was concluded in 1928 by the English physicist Paul A. M. Dirac that such particles were always in motion, even when they seemed to be at rest. The ideal assumption would be to suppose that such particles could be viewed as rotating about an axis. This would give the particle a certain angular momentum.

The particle must also possess energy if it is spinning and energy can be absorbed only a quantum at a time. This is true for all spinning objects, of course, even for a planet like the Earth. The size of a quantum is so small, however, compared to the total energy contained in the Earth's motion that if the Earth were to gain a quantum of rotational energy or even a trillion quanta, no one could tell the difference. If a subatomic particle were to gain such a quantum of energy, however, its spin would change considerably, for a quantum is quite large compared to a single subatomic particle. For that reason, particle spin cannot be represented by just any measurement at all, but can be shown to possess only those values that would represent whole numbers of quanta of energy.

The actual size of the angular momentum possessed by a spinning particle is extremely small in terms of ordinary units, and so a special scale was devised, in which the spin of the photon was set equal to 1 exactly. On that scale, the proton and electron each have a spin of $\frac{1}{2}$.

Angular momentum can, however, exist in two opposed varieties (see page 18), clockwise and counterclockwise. A proton or electron could spin either way, and the spin can therefore be represented as either $+\frac{1}{2}$ or $-\frac{1}{2}$.

Suppose, now, we had a system containing a number of such particles. If the law of conservation of angular momentum is to remain valid, the total spin of the system ought to be equal to the sum of the spins of the individual particles.

Consider a system of four particles, either protons or electrons or a mixture of both. If each particle has a spin of either $+\frac{1}{2}$ or $-\frac{1}{2}$, then the total spin must be either 0 or an integer. Thus, if the spins are $+\frac{1}{2}$, $+\frac{1}{2}$, $-\frac{1}{2}$, and $-\frac{1}{2}$, the sum is 0; if they are $+\frac{1}{2}$, $-\frac{1}{2}$, $-\frac{1}{2}$, and $-\frac{1}{2}$, the sum is -1; and if they are $+\frac{1}{2}$, $+\frac{1}{2}$, $+\frac{1}{2}$, and $+\frac{1}{2}$, the sum is $+2$, etc.

This is true not only for a system of four particles, but for any system containing an even number of particles where each particle, individually, has a spin of $+\frac{1}{2}$ or $-\frac{1}{2}$. The total is always either 0 or an integer.

Suppose, on the other hand, you were dealing with three particles. The spins might be $+\frac{1}{2}$, $+\frac{1}{2}$, and $-\frac{1}{2}$, for a total of $+\frac{1}{2}$; or $-\frac{1}{2}$, $-\frac{1}{2}$, and $-\frac{1}{2}$, for a total of $-\frac{3}{2}$. However you distributed the positive and negative signs, the sum would always be equal to a "half-integer" and never either to 0 or an integer. This is true for any system containing an odd number of particles, where each particle, individually, has a spin of $+\frac{1}{2}$ or $-\frac{1}{2}$.

If atomic nuclei consist of protons and electrons, the net spin of the nucleus (*nuclear spin*) must therefore depend on the total number of such particles present. If the nitrogen nucleus, $_7N^{14}$, is indeed composed of 14 protons and 7 electrons, the total number of particles it contains is 21, an odd number, and the nuclear spin of nitrogen-14 should be a half-integer. Experiments in 1929 showed, however, that it is not; it is an integer.

This is true in the case of certain other nuclei too, so that it became abundantly clear that if the nucleus contained both protons and electrons, several of them violated the law of conservation of angular momentum. Physicists hated to abandon the law if they could possibly avoid doing so, and cast about for some other explanation of nuclear structure.

One suggestion was that in place of a proton-electron combination, there might be present a single uncharged particle. As far as conservation of electric charge was concerned, this change would make no difference. A proton and electron, taken together, would have a charge of $+1$ plus -1, or 0, and a single uncharged particle would also have a charge of 0.

There could well be a difference in angular momentum, however. If a proton and electron each has a spin of either $+\frac{1}{2}$ or $-\frac{1}{2}$, then together they can have an overall spin of $+1$, 0, or -1. The two together could never have an overall spin of $+\frac{1}{2}$ or $-\frac{1}{2}$. A single uncharged particle might, however, very well possess a spin of $+\frac{1}{2}$ or $-\frac{1}{2}$.

And in that case, what a difference. The nitrogen-14 nucleus might then consist of 7 protons plus 7 uncharged particles. If the uncharged particles were as massive as protons, the mass number would then be 14, while the atomic number (based on the protons alone since only they were positively charged) would be 7. We would still have the isotope $_7N^{14}$. Only now the total number of particles in the nucleus would be 14, an even number, instead of 21, an odd number.

With an even number of particles, each with half-spin, the spin of the nitrogen nucleus can be expected to have a value that is an integer. The law of conservation of angular momentum would be saved.

The difficulty lay in translating theory into demonstration, in actually detecting the existence of an uncharged particle.

All the methods for detecting subatomic particles rested upon the ability of those particles to knock electrons out of atoms with which they collided and thus form ions. It is the presence of these ions that are detected by the various devices used by physicists to study particles (see page 51).

Ions can be formed by particles carrying either type of charge. A negatively charged particle repels the negatively charged electrons and pushes them out of any atom it approaches closely. A positively charged particle attracts electrons and pulls them out of the atoms it approaches. An uncharged particle does neither, does not form ions, therefore, and cannot be detected directly.

Nevertheless, there are indirect methods for detecting objects ordinarily invisible. If you look out a window, you might see trees but you won't see air. However, if you note the branches of the trees swaying, you can reasonably assume that they are gaining energy at the expense of some moving mass you cannot see. By carefully studying the behavior of moving branches, you could learn a great deal about the properties of air without ever seeing it.

Beginning in 1930, it was discovered that when certain elements were exposed to alpha-particle bombardment, a radiation was emitted which could not be detected by the usual methods. How did one know it was there, then? Because if paraffin were placed in the path of such radiation, protons were hurled out. This was like the wind moving the trees; something had to be imparting momentum to the protons. The momentum imparted was considerable so that the radiation had to consist of particles that were quite massive, quite rapidly moving, or both.

The English physicist James Chadwick interpreted the evidence properly in 1932, and announced the discovery of the long-suspected neutral particle. It was named the *neutron*. The neutron has a mass that is just a shade over that of the proton; 1.008665 is the currently accepted figures. It has an electric charge of 0 and a spin of $+\frac{1}{2}$ or $-\frac{1}{2}$, exactly the set of properties required to save the law of conservation of angular momentum.

It was at once suggested by the German physicist Werner Karl Heisenberg that the nucleus be considered as made up of protons and neutrons. (It is these two which are the two varieties of nucleons mentioned on page 44. In other words, a nucleon can be either a proton or a neutron.)

Since protons and neutrons both have a mass number of just about 1, the mass number of any nucleus is equal to the sum of the nucleons it contains. The atomic number, representing the electric charge of the nucleus, is the sum of the protons alone, since only the protons contribute to that charge. Thus, the helium-4 nucleus, $_2He^4$ is composed of 2 protons and 2 neutrons, for a total of 4 nucleons. Oxygen-16, $_8O^{16}$, is made up of 8 protons and 8 neutrons, for a total of 16 nucleons. Thorium-232, $_{90}Th^{232}$, is made up of 90 protons and 142 neutrons for a total of 232 nucleons.

Since all the isotopes of a particular element are equal in atomic number, all the isotopes must have the same characteristic number of protons in the nucleus. Since their mass numbers differ, they must have a different total number of nucleons and this can arise only from differences in the number of neutrons. Thus, the two carbon isotopes, carbon-12 ($_6C^{12}$) and carbon-13 ($_6C^{13}$) are made up of nuclei containing 6 protons and 6 neutrons in the first case and 6 protons and 7 neutrons in the second.

As for uranium, uranium-235 ($_{92}U^{235}$) is made up of 92 protons and 143 neutrons for 235 nucleons altogether, while uranium-238 ($_{92}$U^{238}) is made up of 92 protons and 146 neutrons for 238 nucleons altogether.

NEUTRON BREAKDOWN

The proton-neutron model of the nucleus has proved quite satisfactory and it is still accepted today as the best expression of the basic nuclear structure. Nevertheless, it raises questions. If the atomic nucleus contains only protons and neutrons we must return to the problem of how negatively charged electrons can be hurled out of it in the form of beta particles.

The fair conclusion seems to be that if the electron is not present in the nucleus to begin with, it must be formed at the moment it is hurled out. If so, we can use the various conservation laws to guide us in working out the events. If an electron is formed, the negative electric charge upon it must also be formed. By the law of conservation of electric charge, such a negative charge cannot be formed unless a positive charge is simultaneously formed. No particle carrying a positive charge is hurled out of the nucleus along with the beta particle; therefore such a particle must remain inside the nucleus. The only positively charged particle known to exist inside the nucleus is the proton.

We can therefore conclude that when an electron is hurled out of a nucleus, a proton is formed inside the nucleus.

This takes care of the law of conservation of electric charge, but brings us face to face with the law of conservation of energy. A proton possesses mass and if it is formed it must be at the expense of disappearance of mass elsewhere. As it happens, there are neutrons present in all nuclei but that of hydrogen-1. A neutron, being uncharged, can appear or disappear without affecting the law of conservation of electric charge. Therefore, we can say that when a beta particle is hurled out of the nucleus, a neutron disappears and a proton appears at the same time within the nucleus. (See Figure 8.)

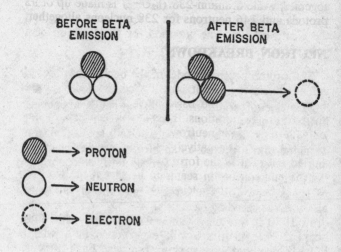

FIGURE 8. Beta-Particle Emission

The interconnection is more obvious if we say that a neutron is converted into a proton and ejects an electron at the same time. The law of conservation of energy can explain this in fine detail for the neutron is slightly more massive than the proton. The proton and electron together have a mass of 1.008374 on the atomic weight scale, while the neutron has a mass of 1.008665. When the neutron is converted to an electron plus a proton, a mass of 0.00029 "disappears." Actually it is converted into energy, about 320 Kev of it, and this is available to appear as the kinetic energy of the speeding beta particle.

This seems satisfactory, so let us summarize by placing it into as uncomplicated a system of symbols as possible. From now on, in this book, we will be concerned almost entirely with subatomic particles. As far as electric charge is concerned these will come in three varieties; those possessing charges of 0, +1, and —1. Let us indicate the charges by a mere + or —, or nothing at all for the uncharged particle. We can then symbolize the neutron as n, the proton as p^+ and the electron as e^-. The equation for the emission of a beta particle can then be written:

$$n \longrightarrow p^+ + e^-$$

This line of argument indicates what is happening inside the nucleus in only an indirect way. One can't actually look inside the nucleus and see a neutron changing into a proton as an electron comes charging out. Not yet, at least. But suppose one could observe individual neutrons in the free state. Would they, so to speak, convert into protons in front of our eyes and liberate speeding electrons?

In 1950 it was finally shown that exactly this does happen. Free neutrons will, every once in a while, break down and change into protons. Each time a neutron undergoes such a change, an electron is emitted.

I say "every once in a while." It does not happen in a flash. Free neutrons will exist as free neutrons a period of time before undergoing breakdown, and the question of how long that period of time might be brings up an important point.

It is impossible to tell when any particular neutron will undergo radioactive breakdown. For individual particles, it is purely a matter of chance. One might last a millionth of a second; another five weeks; still another twenty-seven billion years. If, however, a great many particles of a particular type are taken into account, it is possible to foretell with considerable accuracy how long it will take a certain percentage to break down.

(This is analogous to the manner in which insurance actuaries can predict longevity. While the insurance statistician cannot predict how long a particular man will live, he can, given a large group of men of a particular age, occupation, place of abode, etc., predict with considerable accuracy how long it will be before half of them are dead.)

It is customary to speak of the period over which half the particles of a given type break down as the *half-life* of that particle (a term introduced by Rutherford in 1904). Each different kind of particle has its own characteristic half-life.

The half-life of uranium-238, for instance, is 4,500,-000,000 or 4.5×10^9 years. The half-life of thorium-232 is even longer; 14,000,000,000 or 1.4×10^{10} years. It is for this reason that uranium and thorium are still found in considerable quantities in the Earth's crust despite the fact that some of their atoms are breaking down at every given moment. Through all of Earth's five-billion-year history, only half of its uranium-238 supply has broken down and far less than half of its thorium-232 supply.

Some radioactive nuclei are far less stable than this, however. For instance, when uranium-238 emits an alpha particle, it is converted into thorium-234. The half-life of thorium-234 is a mere twenty-four days. That is why there are only traces of thorium-234 in the Earth's crust. It forms very slowly from uranium-238 and what is formed breaks down very quickly.

Thorium-234, in breaking down, emits a beta particle. This means that, within its nucleus, a neutron is changing into a proton. Such neutron/proton conversions in a large quantity of thorium-234 atoms are proceeding at a rate sufficient to yield a half-life of twenty-four days. In some radioactive isotopes, neutrons are converted into protons at much slower rates. For instance, potassium-40 emits beta particles with a half-life of 1,300,000,000 or 1.3×10^9 years.

In some isotopes, no radioactive breakdown takes place at all. In the nuclei of the atoms of oxygen-16, for instance, no neutron ever turns into a proton of its own accord, as far as we know. The half-life is infinite in length.

What is most interesting for our present purposes, however, is the half-life of a free neutron, a neutron that exists by itself and not as part of an atom. A free neutron is not surrounded by other particles whose influence may render it more stable or less stable and therefore lengthen or shorten its half-life. With a free neutron, we get an untouched half-life, so to speak, and it turns out to be just about twelve minutes.

Of a trillion neutrons, let us say, half will be converted into protons—emitting electrons in the process —by the end of twelve minutes. Not very long!

6

ANTIPARTICLES

LEPTONS AND BARYONS

Let's stop long enough, now, to cast a glance over the different subatomic particles we have been discussing. First, there are the nuclei of the various elements. These, however, can be dismissed, for all atomic nuclei but that of hydrogen-1 are made up of still smaller particles, and it is those still smaller particles we are interested in.

Nuclear physicists are particularly interested in particles that can't be broken down into smaller ones and which therefore represent a kind of minimum size. These are the *elementary particles.**

I have mentioned four particles so far that may be elementary: the proton, the neutron, the electron, and the photon. These may be divided into two groups. The proton and neutron are relatively massive. They, and other massive elementary particles discovered since 1932, are all lumped together as *baryons,* from a Greek word meaning "heavy."

The electron and photon, on the other hand, have little mass (the photon is massless, in fact). They, and other light particles discovered since 1932, are lumped

---

*By the middle of the twentieth century, physicists weren't at all sure that they knew just what an elementary particle might be. It is possible that particles which are usually considered to be elementary may be made up of still simpler particles after all. This need not concern us in this book, however.

together as *leptons,* from a Greek word meaning "small" or "weak."

The four elementary particles can be divided in another fashion. Three of them, the proton, electron, and photon, are *stable.* In other words, if a single proton (or electron, or photon) were alone in the universe, it would remain unchanged forever, as far as we know. (To be sure, any of these three may be changed into other forms through interaction with other particles, but no change takes place without such interaction.)

The neutron, on the other hand, is *unstable.* If a single neutron existed alone in the universe, it would, sooner or later, very probably in a mere matter of minutes, break down and become a proton and an electron. Such instability is inherent in the nature of the particle itself and does not depend on the presence of other kinds of particles.

Why should the neutron differ from the other three particles by being unstable? The conversion of a neutron to a proton and an electron involves a loss of mass. It is this loss of mass, apparently, that is critical. It turns out that in every spontaneous breakdown, there is a loss of mass. The mass lost is, of course, converted into energy, so we might say there is a general tendency in the universe to pass from matter to energy.

That explains at once why the photon is stable. It has zero rest-mass and therefore can't possibly break down into any particle of less mass. It is stable for default of anything to break down into. By the same line of reasoning, any massless particle is stable.

This does not, however, explain the electron's stability. The electron has a rest-mass that is tiny but not zero, and if the universal tendency is to change mass into energy, why is the electron spared? Why is it not broken down into one or more photons of equivalent mass, but zero rest-mass (see page 62)?

That this doesn't happen can be considered an expression of the law of conservation of electric charge. A photon carries no electric charge and if an electron broke down into one or more photons, what would become of the electron's negative electric charge? As far as physicists now know, there is no particle that is less massive than the electron and still capable of carrying a negative electric charge. It is for that reason that the electron does not break down; there is simply no way for it to dispose of its charge if it does.

And the proton? One might suppose that it could be the least massive possible particle capable of carrying a positive charge. If so, it would be stable for the same reason that the electron is. This explanation, however, will not hold up. The proton, as it turned out, is not the least massive particle capable of carrying a positive charge.

POSITRONS

The first indication of a positively charged particle less massive than a proton came in 1930, when Paul Dirac worked out a mathematical treatment of certain properties of the electron. He found that if his mathematics was to be trusted, an electron ought to exist in two different forms. In one of those forms, it was the ordinary well-known electron which had already been studied, at that time, for over thirty years. In the other form, it was just like an ordinary electron in every way, except that it carried a positive charge instead of a negative one.

Two years later, in 1932, an American physicist, Carl David Anderson, was studying cosmic rays, which are an intensely energetic form of radiation bombarding Earth from outer space. He found evidence in his particle-detecting devices of something that behaved exactly like an electron except that under the influence of a magnet's pull it curved in the wrong direction. This meant it carried a positive charge instead of a negative one. Anderson had discovered Dirac's positively charged electron.

Anderson gave the particle the name of *positron,* because of its positive charge and that remains the most-used title for the particle. This particular name, however, is unfortunate in that it tends to obscure its close relationship to the electron.

Sometimes the electron and positron are termed the *negative electron* and *positive electron*, respectively, and this is reflected in the symbols commonly used for the particles. Since the electron is e^-, the positron is very often represented as e^+.

Another device is to let the electron keep its name but to call the positron an *antielectron*, wherein the prefix "anti-" carries the notion of "opposite." In many ways antielectron is the best name of all, for other particles have their opposites and all of these but the positron make use of the prefix "anti-." In fact these opposite particles are lumped together as *antiparticles*.

It is now customary to represent an antiparticle by the symbol of the particle with a horizontal line added above. Thus, a positron can be represented as $\overline{e^+}$ to indicate that it is not merely a positively charged electron, but that it is an antiparticle as well.

Soon after Anderson had discovered the positron, it was found to be produced by certain radioactive atomic nuclei—not by nuclei to be found in nature, to be sure, but by some formed in the laboratory.

The first such nucleus to be discovered was phosphorus-30. This was formed in 1934 by a husband-and-wife team of French scientists, Frédéric and Irène Joliot-Curie, who obtained it by bombarding aluminum atoms with alpha particles. They found that phosphorus-30 emitted positrons spontaneously (as *positive beta particles,* so to speak) and that silicon-30 was produced in the process. Since phosphorus has an atomic number of 15 and silicon one of 14, the radioactive change could be represented thus:

$$P^{+15} \longrightarrow Si^{+14} + \overline{e^+}$$

As you see, electric charge is conserved since $14+1=15$.

You might well wonder what goes on inside a nucleus

to bring about the emission of a positron. If we compare silicon-30 to phosphorus-30, we see that the mass number (30) is identical, so that the total number of nucleons must be the same in both. On the other hand the atomic number of the silicon-30 nucleus (+14) is one less than that of phosphorus-30 (+15). This means that the silicon-30 nucleus must contain one proton less than the phosphorus-30 nucleus. To bring about a decrease of one proton without changing the total number of nucleons, one neutron must simultaneously be added. In other words, a positron is emitted when, within a nucleus, a proton is converted into a neutron, reducing the atomic number by 1 and leaving the mass number unchanged.

This is exactly the reverse of the situation that leads to the emission of an electron (where a neutron is converted to a proton) but this is to be expected. The positron is the antiparticle of the electron, its reverse image, so to speak, and anything the electron can do, the positron can do in reverse, and vice versa.

Another question might now arise. The proton is less massive than the neutron, so it is not surprising that the neutron spontaneously changes into a proton, since spontaneous changes always go in the direction of a decrease in mass. But in that case, how can a proton spontaneously change into a neutron, liberating a positron?

The answer is rather tricky. It is true that a proton is less massive than the neutron when the free particles are considered. Within the nucleus, however, there are energy changes that slightly alter the mass of the individual nucleons. Sometimes a nucleus loses mass if a proton changes to a neutron; and sometimes if a neutron changes to a proton—it depends entirely on the makeup of the nucleus. In the former case, as in phosphorus-30, positrons are emitted; in the latter case, as in thorium-234, electrons are emitted.

Then, of course, there are nuclei possessing a combination of protons and neutrons that are already at minimum mass. A conversion of a proton to a neutron will increase the mass and so will a conversion of a neutron to a proton. In the case of those particular nuclei, neither conversion will take place spontaneously and the nucleus is stable (unless it is one of those combinations that are so massive as to give off alpha particles).

It should be repeated, though, that when neutrons and protons are considered as free and isolated particles, neutrons will spontaneously convert to protons, but never vice versa.

The positron, like the electron, is a stable particle. That is, left to itself, it will never, as far as we know, undergo any change. We can see why that is, since the positron is the least massive particle capable of carrying a positive electric charge. Its stability is an expression of the law of conservation of electric charge.

The positron is not alone in the universe, however. When it is formed, it finds itself in a universe composed of uncounted numbers of other particles, including electrons. Under ordinary conditions, a positron will collide with an electron within a millionth of a second and when particle meets antiparticle, both cease to exist. (It is like a wooden peg encountering a hole on a wooden surface into which it fits exactly; an "antipeg." When peg encounters hole, both cease to exist, and there is an unbroken wooden surface instead.)

In the union of electron and positron, the various conservation laws hold. If the two meet head-on, traveling at identical speeds in opposite direction the two momenta add up to zero. If the electron has a spin of $-\frac{1}{2}$ and the positron one of $+\frac{1}{2}$, the total angular momentum of the system is zero too. Since the electron has a charge of -1 and the positron a charge of $+1$, the total electric charge of the two particles is zero. So far annihilation is complete.

That leaves only energy which, unlike the other quantities, does not exist in positive and negative forms and which cannot therefore be manipulated in such a way as to add up to zero. When electron and positron meet, the energy associated with their mass and their motion must remain in existence in one form or another. The two particles are converted into photons.

As I said earlier (see page 60), the energy equivalence of an electron is 0.51 Mev. Since the positron has the same mass as the electron, the energy equivalence of the two taken together is 1.02 Mev. Therefore, every time a positron-electron pair undergo mutual annihilation, 1.02 Mev of energy must be liberated. This was found to be exactly true experimentally, and this was an excellent reaffirmation of the law of conservation of energy on a subatomic scale.

If we again consider our electron-positron system to have zero momentum, zero angular momentum, and zero charge, what of the photons produced? Photons have no charge so that's all right, but they do have momentum and angular momentum. If one photon is produced, momentum and angular momentum would have to be created, which does not happen.

Instead, two photons are produced, each carrying 0.51 Mev of energy, and emitted in opposite directions. This maintains a net momentum of zero. One photon has a spin of +1, the other of −1, and that maintains a net angular momentum of zero.

If the electron and positron meet in such a fashion as to leave a net momentum or angular momentum of more than zero, that too is maintained. Suppose both particles have a spin of +½ for a total spin of +1. In that case a single photon of +1 spin could be emitted, if the system possessed momentum. If the system did not possess momentum the spin could still be taken care of through the emission of three photons, each with an energy of 0.34 Mev emitted toward the three apices of an equilateral triangle. That would keep the net momentum at zero. If the spins of the three photons were +1, +1, and −1, respectively, the total spin would be maintained at +1.

In short, no matter how you arrange electron-positron annihilation, you can end up conserving the four great quantities: momentum, angular momentum, electric charge, and energy.

It works the other way, too; energy can be converted into mass. A gamma-ray photon of 1.02 Mev can, under certain conditions, disappear and be replaced by an electron-positron pair. A less energetic photon lacks the energy required and a more energetic one adds the additional energy in the form of the kinetic energy of the moving particles.

A gamma-ray photon cannot be converted into an electron alone or into a positron alone. If either could be done, electric charge would have to be created and the law of conservation of charge would be broken.

We can represent the interrelationship between an electron-positron pair and gamma-ray photons as follows, if we symbolize the gamma-ray photon with the Greek letter "gamma" (γ):

$$e^- + \overline{e^+} \longrightarrow \gamma + \gamma$$

ANTINUCLEONS

Dirac's theory, which predicted the existence of a positively charged electron, could also be made to apply to the proton. There should, therefore, be a negatively charged proton; an *antiproton*. Once the positron was discovered, physicists were convinced the antiproton also existed, or could be made to exist.

The difficulty was that a proton is 1836 times as massive as an electron. A minimum energy of 1.02 Mev sufficed to create an electron-positron pair and it followed that 1872 Mev of energy (1.872 Bev) would be the absolute minimum required to create a proton and antiproton. It was not until the 1950s that physicists produced devices capable of concentrating energy to that extent. In 1956 the Italian-American physicist Emilio Segrè and his American colleague Owen Chamberlain accomplished the task, brought about the formation of antiprotons, and demonstrated their existence conclusively.

Protons and antiprotons can undergo mutual annihilation when they meet just as electrons and positrons can. All I have said about the latter situation applies to the former except, of course, that in the case of the protons and antiprotons, far greater energies are involved. If we symbolize the antiproton as p^-, we can write:

$$p^+ + \overline{p^-} \longrightarrow \gamma + \gamma$$

In 1965 the reverse was achieved. High-energy gamma rays were converted into proton-antiproton pairs.

In one respect, the interaction of proton and an antiproton introduces something new. Almost as soon as the antiproton was discovered it was found that occasionally, if a proton and antiproton scored a near-miss, the electric charge on both was canceled but mass was not annihilated. Where two charged massive particles had existed before, two uncharged massive particles existed afterward. In place of the proton was a neutron and in place of the antiproton was an *antineutron*. If the latter is symbolized $\overline{n^-}$, we can express this as follows:

$$p^+ + \overline{p^-} \longrightarrow n + \overline{n}$$

Just as the neutron and proton are lumped together as nucleons, the antineutron and antiproton can be considered as *antinucleons*. Both nucleons and antinucleons are included among the baryons (just as the positron is included among the leptons).

But what can an antineutron be? The positron differs from the electron in its charge, and the antiproton differs from the proton in its charge. The neutron, however, is uncharged, and so is the antineutron. In what way, then, are they different? How are they opposites?

The answer, apparently, lies in the matter of spin. Suppose you consider a subatomic particle to be a tiny spinning sphere, turning on its axis and possessing two poles. If you view the particle from above one pole, it seems to be spinning counterclockwise; from above the other pole, it is spinning clockwise. Let us call the pole from which a counterclockwise spin is viewed as the "north pole." (This is analogous to the situation on Earth, for if our planet is viewed from above its north pole, its rotation from west to east makes it appear to be turning in the counterclockwise sense.)

If a spinning particle carries a charge, it sets up a magnetic field as it spins, and there is a north magnetic pole and a south magnetic pole. In the proton, the north magnetic pole is on top and coincides with the north pole. In the antiproton, however, the north magnetic pole is on bottom and coincides with the south pole. In other words, the magnetic field of the antiproton is reversed as compared with that of the proton (see Figure 9). In magnetic properties as well as in charge, the two particles are opposites. This is true of any particle/antiparticle pair.

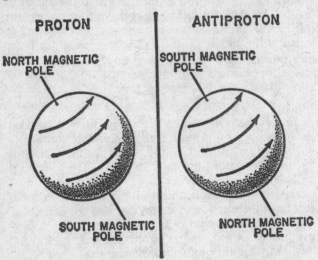

FIGURE 9. Proton and Antiproton Magnetic Poles

Although the neutron has no electric charge, it does have a magnetic field associated with it. The reason for this is not entirely clear, but physicists suspect that although the neutron has an overall electric charge of zero, it has regions of positive charge and of negative charge, which are not distributed symmetrically and this gives rise to the magnetic field.

The neutron, then, has its magnetic field oriented in one direction while the antineutron has it oriented in the other. It is here that the oppositeness is to be found.

The antineutron has precisely the same mass as the neutron and the antiproton precisely the same mass as the proton. This means that the antineutron is somewhat more massive than the antiproton and can break down to one. In the breakdown of an antineutron (with a charge of 0) to an antiproton (with a charge of −1), a negative charge is formed. This cannot take place, if the law of conservation of electric charge is to hold, unless a positive charge is simultaneously created. The positive charge appears in the form of a positron so that:

$$\bar{n} \longrightarrow \bar{p}^- + \bar{e}^+$$

The breakdown of the antineutron is analogous to the breakdown of the neutron (except for the reversal of charge relationship) in every way. Even the half-life is exactly the same in both cases.

The reverse of antineutron breakdown does not take place. The antiproton, left entirely to itself, is stable, just as the proton is, and will (to the best of our knowledge) never alter.

CONSERVATION OF BARYON NUMBER

But this brings us back to the question I asked earlier: Why is a proton stable? We can add another question to that now: Why is an antiproton stable?

It is clearly improper to say that a proton is the smallest mass with which a positive charge can be associated, as we might have been tempted to say before the antiparticles were introduced. The proton could, after all, break down to a positron and gamma-ray photons. Electric charge would be conserved and all the other conservation laws could be made to hold also. Similarly an antiproton could break down to an electron and gamma-ray photons.

It is the experience of scientists that in the subatomic world any event which is not forbidden is compulsory. If some breakdown violates none of the conservation laws, then it must take place. It may take place only very infrequently, but take place it will. If, on the other hand, some subatomic event persistently refuses ever to happen, then some conservation law would be broken if it did.

Since a proton never breaks down to positron, some conservation law must stand in the way. Yet the four conservation laws already discussed do not stand in the way. A conversion of a proton to a positron, both carrying a unit positive charge, would not break the law of conservation of electric charge; and the properties of the positron and photons formed from the proton could easily be so arranged as not to violate conservation of angular momentum, linear momentum, and energy.

The conclusion forced upon the physicist, then, is that a fifth conservation law, one he had not previously considered, is involved.

As psysicists looked through all the subatomic interactions they had studied, it began to appear that a baryon never disappeared altogether. Whenever a baryon of one kind ceased to exist, a baryon of another kind simultaneously came into existence.

Of course when a baryon met its opposite number (as when a proton met an antiproton) both might cease to exist without the production of any other baryon to make up for that disappearance.

To straighten out the matter of these antiparticles, the various subatomic particles were assigned specific *baryon numbers*. The proton and the neutron were each given a baryon number of +1, while the antiproton and the antineutron were each given a baryon number of —1. The various leptons (the electron, positron, and photon) were given baryon numbers of 0.

Now physicists could speak of the *law of conservation of baryon number* which could be stated as follows: *The net baryon number of a closed system is constant.*

(Until now, all the conservation laws I have discussed were first worked out in the everyday world and later applied to the atom. Now we come for the first time to a conservation law which was worked out directly from subatomic events. It will not be the last time.)

Let's consider some examples. In radioactive transformations, a uranium-238 nucleus can break down to a thorium-234 nucleus and an alpha particle (helium-4). Since a uranium 2-38 nucleus contains a total of 238 protons and neutrons, its baryon number is 238. Similarly, the baryon number of thorium-234 is 234 and of the alpha particle is 4. Since $238=234+4$, baryon number is conserved here.

Again, a thorium-234 nucleus can give off a beta particle (which is an electron and therefore has a baryon number of 0) to form protactinium-234. Since $234=234+0$, baryon number is again conserved. It is, indeed, conserved in all known radioactive transformations.

What if we deal with single particles? If a neutron breaks down to a proton and electron, baryon number is conserved since $1=1+0$. If an antineutron breaks down to an antiproton and a position, baryon number is again conserved for $-1=-1+0$.

If a proton and an antiproton interact to form a neutron and antineutron, we find that $1-1=1-1$. And if a proton and antiproton interact to form two gamma-ray photons (or any number of them), we find that $1-1=0+0$.

Indeed, in all known atomic and subatomic events, baryon number is conserved. Physicists have never found an exception.

That tells us why a proton cannot spontaneously decay to a positron, or an antiproton to an electron.

In the first case a baryon number of $+1$ would become 0, and in the latter case, a baryon number of -1 would become 0. Neither change is possible without a violation of the law of conservation of baryon number.

Indeed, the proton and the antiproton are, as far as we know, the least massive baryons capable of existing. This is what keeps them stable. To undergo a spontaneous change of any sort would mean the production of less massive particles; but any less massive particle is not a baryon, and the law of conservation of baryon number therefore prevents this from happening.

This brings up one oddity that should be mentioned. The law of conservation of electric charge would make it seem that no electron can be created out of energy without the simultaneous creation of a positron. The same law, together with the law of conservation of baryon number, would make it seem that no proton could be created without the simultaneous creation of an antiproton.

And yet, in the universe about us, electrons and protons are extremely common, while positrons and antiprotons are extremely rare. Why is that?

No satisfactory answer can yet be supplied. One possibility is that when the universe was created (by whatever means) particles and antiparticles were formed in equal numbers but were somehow separated. Perhaps in addition to our own universe, there exists also an *antiuniverse*. Our universe is made up of ordinary matter, consisting of atoms with nuclei built up of protons and neutrons while electrons fill the outskirts. In the antiuniverse, there would be *antimatter,* consisting of atoms with nuclei built up of antiprotons and antineutrons while positrons fill the outskirts. In the antiuniverse, ordinary matter would be extremely rare.

(Until recently, antimatter remained a mere theoretical concept. In 1965, however, physicists at Brookhaven National Laboratory in Long Island succeeded in producing a short-lived nucleus consisting of an antiproton and an antineutron. As it happens, the nucleus of hydrogen-2 consists of a proton and neutron. Since hydrogen-2 is often called *deuterium,* the proton/ neutron combination is called a *deuteron.* The antiproton/antineutron combination is therefore an *antideuteron.* The antideuteron is the simplest form of antimatter that can exist above the level of the individual particle. Undoubtedly, in time to come, more complex forms of antimatter will be formed in the laboratory.)

It might also be that both matter and antimatter are present in our single universe, but are collected into separate galaxies. It would be difficult to prove whether or not some of the galaxies visible to our telescopes were antigalaxies.

One might speculate that we could tell galaxies from antigalaxies by the light they emit. If ordinary matter produces radiation consisting of photons, might not antimatter produce radiation consisting of "antiphotons"? Might not they be distinguished in this fashion?

Unfortunately not! If antiphotons existed, then the mutual annihilation of particles and antiparticles ought to produce photons and antiphotons in equal numbers. Only photons appear, however, and physicists have therefore concluded that a photon serves as its own antiparticle. The radiation emitted by matter and antimatter would therefore be identical in all respects, and the two could not be distinguished in this matter. (However all hope is not lost, as we shall discover later on.)

If matter and antimatter both exist in our single universe, then it might be that considerable quantities of the two would, on occasion, meet. If so there would be an unusual flood of energy arising from mutual annihilation, energies much greater than would arise from the type of nuclear reactions that proceed in the interior of stars such as our Sun.

There are, indeed, galaxies and other cosmic objects which do produce unusually high floods of energy in the form of either light or radio waves, or both. Astronomers are currently engaged in trying to determine the source of that energy. Matter/antimatter annihilation is one possibility, but not the only one.

At this halfway point in the book let us list the elementary particles we have so far mentioned.

First there are three leptons. Two of these are the electron and its opposite number the positron (or antielectron). The third is the photon.

Then there are the four baryons, which come in two pairs: the proton and antiproton; and the neutron and antineutron.

7

ENTER THE NEUTRINO

ALPHA PARTICLES AND ENERGY

So far the conservation laws have triumphantly met those challenges described in the earlier chapters. Where one of the laws proved to be inadequate, a better interpretation was evolved, so that the old law of conservation of mass, for instance, came to be incorporated into a broader and more effective law of conservation of energy. Again, when events did not happen that might be expected to happen, a new conservation law could be designed, as in the case of the law of conservation of baryon number.

Not all challenges to the conservation laws were easily met, however. A particularly puzzling situation arose quite early in the game in connection with the kinetic energy of particles emitted by radioactive substances.

Ideally, the energy of an alpha particle, for instance, can be determined by measuring the mass of the original radioactive nucleus, that of the alpha particle and of the final nucleus. The alpha particle and the final nucleus should have a combined mass slightly smaller than that of the original nucleus, and the energy equivalence of the missing mass should be equal to the kinetic energy of the alpha particle.

To test this, physicists must be able to measure the mass of various nuclei and other particles with considerable precision, and it was only in the 1920s that this became possible. However, even without knowing the exact masses, certain important conclusions about particle energies could be drawn.

Let us consider thorium-232, for instance, which breaks down into an alpha particle (helium-4) and radium-228. All thorium-232 nuclei have identical mass. All radium-228 nuclei have identical mass also, and so do all alpha particles. Without knowing what those masses might be, we can still say that in every case in which a thorium-232 atom emits an alpha particle, the mass deficit should be the same and the kinetic energy of the alpha particle should, therefore, be the same. Ideally, thorium-232 ought to emit alpha particles with only a single set of energies.

How can one determine the kinetic energy of alpha particles? One way is to measure the distance they can penetrate matter; the more energetic they are, the farther they can penetrate. Alpha particles of any energy content are not very penetrating and are stopped by very thin films of solid matter. However, they can penetrate several centimeters of air. As they pass through air, they are perpetually transferring energy to the molecules of air with which they collide. They slow down, bit by bit, and eventually pick up electrons and become ordinary helium atoms. As helium atoms they can no longer be detected by the methods that suffice for alpha particles and, to all intents and purposes, they have disappeared.

As an example, alpha particles can be detected by a film of a chemical called zinc sulfide, for every time an alpha particle hits the zinc sulfide, it produces a tiny scintillation of light. If a *scintillation counter* is placed near an alpha-particle source (say, a bit of thorium-232 in a lead container with a hole, so that a tight stream of alpha particles issues from the hole) one can determine the number of alpha particles produced per second by the number of scintillations per second.

As the scintillation counter is placed farther and farther from the alpha-particle source, the alpha particles must travel through greater and greater thicknesses of air to reach the counter. If alpha particles are emitted with a variety of energies, then the least energetic ones would quickly drop out, slightly more energetic ones would drop out only after penetrating a somewhat greater thickness of air, still more energetic ones only after penetrating a still greater thickness of air and so on. The result would be that as the scintillation counter was placed farther and farther away, the alpha-particle count would gradually decrease.

If, however, all the alpha particles were given off with substantially equal energies, then all would penetrate the same thickness of air. The scintillation counter, as it was moved away, would count the same number of alpha particles until some key point was reached and then, quite suddenly, it would count none.

It was the latter that was observed to happen by the English physicist William Henry Bragg in 1904. Almost all the alpha particles produced by a particular variety of nucleus had the same energy and showed the same penetration. The alpha particles from thorium-232, for instance, could all penetrate 2.8 centimeters of air. The alpha particles produced by radium-226 can all penetrate 3.3 centimeters of air, and those of polonium-212 can all penetrate 8.6 centimeters of air.*

Actually, though, there are some irregularities. It was discovered in 1929 that small quantities of alpha particles emitted by a given radioactive nucleus might possess unusually high kinetic energies and be more penetrating than the rest. The reason for this was that the original radioactive nucleus could exist in a number of different *excited states*. In those excited states it had a larger energy content than in its normal *ground state*. When a nucleus emits an alpha particle while in

---

*The more penetrating the alpha particles of a particular nucleus, the greater the mass deficit in the course of the radioactive breakdown and the greater the probability of that particular breakdown taking place. In short, the more penetrating the alpha particle, the smaller the half-life. Whereas thorium-232 has a half-life of fourteen billion years, radium-226 has one of 1620 years and polonium-212 has one of three ten-millionths of a second.

an excited state, the alpha particle gains the extra energy. As a result, in addition to the main group of alpha particles, there would be small groups of more penetrating ones, one group for each of the excited states.

Sometimes, when the radioactive nucleus is formed by the breakdown of an earlier nucleus, it exists in the excited state as it forms. In that case, most of the alpha particles it emits are unusually high in energy, with small groups of lower energy also present.

These separate groups of alpha particles of different energies and penetrations (from 2 to as many as 13) are referred to as the *alpha-particle spectrum* of a particular nucleus.

Each component of this spectrum matched what was expected of some particular excited state. No energy was left unaccounted for. As far as alpha particles were concerned, all was well with the law of conservation of energy from the beginning, and all remains well to the present moment.

But not so with beta particles.

BETA PARTICLES AND ENERGY

All the reasoning that applies to the alpha particles should apply to the beta particles as well. When a number of nuclei of a particular variety break down and emit beta particles, the same energy relationships should prevail, ideally, in every case and all the beta particles produced should possess one particular amount of kinetic energy. Or, if the original nucleus can exist in a small number of excited states, the beta particles ought to possess a small number of different kinetic energies.

However, as early as 1900, it began to seem that beta particles could be given off with any amount of energy, up to some maximum. Over the next fifteen years, the evidence piled up until the matter was certain; beta particles formed a continuous spectrum.

Every type of nucleus that breaks down to emit beta

particles loses a certain amount of mass in the process. This mass decrease should show up as the value of the kinetic energy of the beta particle. And, as it happens, no beta particle produced by any nucleus has ever been observed to possess a kinetic energy greater than the equivalent of the mass decrease. The mass decrease in any radioactive breakdown can thus be set equal to the maximum kinetic energy of the beta particles produced in the course of that breakdown. Thus far, at least, the law of conservation of energy holds.

But by the law of conservation of energy, no beta particle should have a kinetic energy less than that equivalent to the mass decrease. The maximum kinetic energy of the beta particle should also be the minimum kinetic energy—but it is not. Beta particles are very commonly given off with less kinetic energy than was to be expected. In fact, hardly any beta particles ever attained the maximum value which the law of conservation of energy seemed to require.

Some beta particles had slightly less than the maximum value of kinetic energy; some considerably less; some a great deal less. The most common value is about one-third the maximum.

On the whole, then, more than half the energy that should arise out of the mass decrease in those radioactive transformations that involved beta-particle production could not be located.

Through the 1920s many physicists began to suspect that it might be necessary to abandon the law of conservation of energy, at least in connection with those processes involving beta-particle production. This was an alarming prospect since the law was so useful and was adhered to so faithfully in every other respect, but what alternative was there?

In 1931 Wolfgang Pauli suggested an alternative. If the beta particle did not carry off all the energy, it might be because a second particle was produced which carried off the remainder. The energy could be divided among the two particles in any proportions. In some cases, almost all the energy would be assigned to the electron in which case it would have nearly the maximum kinetic energy permitted by the mass decrease. In other cases, almost all the energy would be assigned to the second particle, and then the kinetic energy of the electron would be virtually zero. Where the division was more evenhanded, the electron could have intermediate energy values.

What kind of particle would Pauli's suggestion involve? To reason that out, let's begin by remembering that beta particles are produced whenever a neutron within a nucleus is converted to a proton (see page 76. But if we are to deal with neutron-to-proton conversions, surely it will be simpler to do so if we deal with the free neutron. Pauli didn't have the opportunity to do that when he first advanced his theory, for the neutron had not yet been discovered, but we can take advantage of hindsight.

When a free neutron breaks down to a proton and an electron, the electron can come off with any value of kinetic energy up to a maximum which, in this case, is about 0.78 Mev. This is quite analogous to the situation in which we find ourselves with respect to the beta particles emerging from radioactive nuclei, so that Pauli's particle must be involved in the breakdown of a free neutron.

Let's symbolize Pauli's particle with the common symbol for the unknown, x, just to begin with, and see what properties we can deduce for it. The equation for the breakdown of the neutron can be written:

$$n \longrightarrow p^+ + e^- + x$$

In the first place, if the law of conservation of electric charge is to be upheld in neutron breakdown, x must be uncharged. In that case $0 = 1 - 1 + 0$.

Let's consider mass next. When a neutron breaks down to a proton and an electron, the loss in mass is 0.00029 on the atomic weight scale. This is equivalent to the mass of about half an electron. Therefore, even if all the energy produced by the disappearing mass is pumped into particle x, and if all went to make up mass, x would still have only half the mass of an electron.

We can decide immediately, then, that x must be lighter than an electron. Indeed, it must be considerably lighter, since the electron usually gets a good percentage of the available energy, sometimes nearly all of it. Furthermore, the energy assigned to x is not likely to go entirely into mass; much of it may be the kinetic energy of x.

As the years went by, therefore, the physicists' estimate of the mass of x grew smaller and smaller until finally it became clear that x, like the photon, was a massless particle; that is, it had a rest-mass of zero. Consequently, like the photon, it traveled at the velocity of light from the moment it was formed. Just as photons came in a variety of energies depending on the wavelength associated with them, so x came in a variety of energies, depending on something analogous to wavelength.

Pauli's particle emerges, then, as a massless, uncharged particle and, if so, that would explain why it did its work in so invisible a fashion. Particles were ordinarily detected by the ions they formed and only charged particles formed ions (see page 74). The neutron, an uncharged particle, was detected finally through the effects of its considerable mass. A particle uncharged and also massless deprived the physicist of any reasonable handle by which it might be trapped and studied.

Soon after Pauli's suggestion that x existed, it received a name. Since it was uncharged, the first thought had been to call it a "neutron" but then, the year after Pauli's suggestion, Chadwick discovered the massive uncharged particle that received that name (see page 75). The Italian physicist Enrico Fermi, mindful that x was much less massive than Chadwick's neutron, suggested the name *neutrino* for x. In Italian that name means "little neutral one" and the suggestion was perfect. The name has been used ever since.

The neutrino is usually given, as a symbol, the Greek

letter "nu" (v), so the equation for neutron breakdown may be written:

$$n \longrightarrow p^+ + e^- + v$$

THE NECESSITY OF THE NEUTRINO

Pauli's suggestion of the existence of the neutrino and Fermi's subsequent working out of a detailed theory of neutrino production received a mixed reception from the world of physics.

To be sure, no one wanted to abandon the law of conservation of energy. On the other hand, there was considerable doubt as to whether it was worth saving by means of a particle without mass and without charge; a particle that could not be detected; a particle whose sole reason for existence was merely to save the law. Some physicists considered it only the ghost of a particle, only a gimmick to save the energy "bookkeeping." In fact, one might almost say that the concept of the neutrino was simply a way of saying "the law of conservation of energy doesn't work."†

But it turned out almost at once that the law of conservation of energy was not the only conservation law that was saved by the existence of the neutrino.

Consider a motionless neutron, one with a momentum of zero relative to the observer. When it breaks down, the proton and electron should have a combined linear momentum of zero, if they are the only two bodies formed. The electron should come off in one direction and the proton recoil in exactly the opposite direction

---

†Indeed, if I had succumbed to the temptation to bring in the neutrino very early in this book, I would have had considerable difficulty in proving that this wasn't true, that the neutrino wasn't a figment of scientific mysticism. However, because the first half of this book emphasizes the value and importance of the conservation laws, one can now demonstrate that the neutrino—for all its odd properties—can be accepted as a real particle; even as a completely necessary one.

(but with lower velocity because of its greater mass) very much like a bullet and a rifle.

This, however, is not what happens. The electron and proton are not emitted in exactly opposite directions, but in directions that form a definite angle. The two particles together have a small net momentum in the direction in which they are angled. Momentum is created apparently out of nothing, and the law of conservation of momentum is violated. If a neutrino is also formed, however, it could be expected to come hurtling off in such a direction as to just balance that net momentum of the other two particles (see Figure 10). In other words, with the neutrino (but never without) the law of conservation of momentum is upheld.

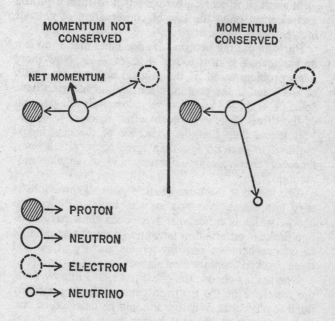

FIGURE 10. Neutron Breakdown

Analogous, and easier to see, is the case of angular momentum. The neutron, proton, and electron each have spins of either $+\frac{1}{2}$ or $-\frac{1}{2}$. Let us suppose that the neutron has a spin of $+\frac{1}{2}$. When it breaks down to a proton and electron, the net spin of those two particles taken together (if they are the only ones formed) ought to equal $+\frac{1}{2}$ if the law of conservation of angular momentum is to be conserved. But is that possible?

The spins of the proton and the electron can be $+\frac{1}{2}$ and $+\frac{1}{2}$; or $+\frac{1}{2}$ and $-\frac{1}{2}$ or $-\frac{1}{2}$ and $-\frac{1}{2}$. The net spins of both particles together are $+1$, 0, and -1 for these cases respectively. It is never $+\frac{1}{2}$ and never can be $+\frac{1}{2}$. If the original spin of the neutron were $-\frac{1}{2}$, the situation is no better, for the net spin is never $-\frac{1}{2}$ and never can be $-\frac{1}{2}$.

In short, if a neutron breaks down to form a proton and electron only, the law of conservation of angular momentum is violated.

But suppose the neutrino is added and that it too can be assigned a spin of either $+\frac{1}{2}$ or $-\frac{1}{2}$. Now there is no problem at all. If the original neutron has a spin of $+\frac{1}{2}$ and if the proton, electron, and neutrino have spins of $+\frac{1}{2}$, $+\frac{1}{2}$, and $-\frac{1}{2}$ respectively the net spin of the three particles taken together is, clearly, $+\frac{1}{2}$.

It seems then that the existence of the neutrino is required to save no less than three conservation laws; those of energy, of momentum, and of angular momentum.

The fact that the same particle does all three jobs is very impressive. A physicist who finds it difficult to choose which is worse, one mysterious ghost particle or one broken conservation law, finds it much less difficult to choose between one mysterious ghost particle and three broken conservation laws.

One had to choose the ghost particle. Little by little, the existence of the neutrino came to be accepted by nuclear physicists. Whether it could be detected or not, they ceased to doubt that it was there. In fact, it was there of necessity. The conservation laws made that plain.

CONSERVATION OF LEPTON NUMBER

The neutrino made it possible not only to save three conservation laws, but to construct a new conservation law as well. To see how that comes about, let's consider the application of the neutrino to antiparticles.

The antineutron breaks down to form an antiproton and a positron (antielectron). The situation is analogous to that of neutron breakdown. The positron comes off with less kinetic energy than it should, the positron and antiproton do not speed away in directly opposite directions, and their spins don't add up properly.

In this case, too, adding a neutrino will balance everything. But in doing so one may ask whether the neutrino produced by the breakdown of an antineutron is the same as that produced by the breakdown of a neutron.

There is the possibility of a difference, you see. The neutrino, like the neutron, is a neutral particle possessed of spin. Like the neutron, the spinning neutrino might set up a magnetic field. If so, this magnetic field can have one of two orientations and in that case we might have a neutrino and an *antineutrino,* just as we have a neutron and antineutron (see page 90).

If this is so, it might seem only natural that the breakdown of a neutron would produce one of these neutrino twins and the breakdown of the antineutron the other. But which neutrino would go with which breakdown? Let's see if we can't answer that.

I have already described the law of conservation of baryon number (see page 93) which states that the net baryon number of a closed system remains constant. Is there an equivalent *law of conservation of lepton number* in which *the net lepton number of a closed system remains constant?* Why shouldn't leptons be taken care of as neatly as baryons are?

Unfortunately, if the neutrino is left out of consideration, they aren't.

The three leptons listed at the end of the previous

chapter are the photon, electron, and positron. Suppose we give the electron a *lepton number* of $+1$, and the positron (which is an antielectron, after all) one of -1. The photon, which is its own antiparticle, can be given neither $+1$ nor -1 as its lepton number and should, most logically, receive a lepton number of 0. All baryons, of course, also have lepton numbers of 0.

Consider the breakdown of the neutron once again now. We begin with a single neutron which has a baryon number of 1, and a lepton number of 0. Suppose that in neutron breakdown only a proton and electron are formed. The proton and electron, taken together, should have a baryon number of 1 and a lepton number of 0, if both are to be conserved. Sure enough, the baryon number of the two together is $1+0$, which adds up to $+1$ and conserves baryon number. The lepton number of the proton and electron, however, is $0+1$ and this also adds up to $+1$. Since in the reaction we begin with a lepton number of 9 and end with one of $+1$, lepton number is not conserved.

Suppose, though, you add the neutrino and antineutrino to the list of leptons, giving the former a lepton number of $+1$, and the latter a lepton number of -1. In that case you need only suppose that the neutron breaks down to a proton, electron, and antineutrino. You begin with a lepton number of 0 and end with a lepton number of $0+1-1=0$. Lepton number is conserved and you can represent neutron breakdown as:

$$n \longrightarrow p^+ + e^- + \bar{\nu}$$

where, of course, $\bar{\nu}$, represents the antineutrino.

By the same reasoning, when an antineutron (lepton number, 0) breaks down, it must form an antiproton, a positron, and a neutrino. The lepton numbers of the three particles produced are 0, -1, and $+1$ respectively and this adds up to zero:

$$\bar{n} \longrightarrow \overline{p^-} + \overline{e^+} + \nu$$

In the free state, neutrons and antineutrons break down to protons and antiprotons respectively but the reverse situation does not take place. Within nuclei, however, it is sometimes possible for protons to be converted spontaneously to neutrons, as in the case of phosphorus-30 (see page 84). By analogy, in antimatter, it is possible for antiprotons to change to protons.

When a proton is converted to a neutron, a positron and neutrino are also formed:

$$p^+ \longrightarrow n + \overline{e^+} + \nu$$

When an antiproton is converted to an antineutron, an electron and antineutrino are also formed:

$$\overline{p^-} \longrightarrow \overline{n} + e^- + \overline{\nu}$$

In both cases, lepton number is conserved, for $0=0 -1+1$ in the first case and $0=0+1-1$ in the second. We can summarize by saying that when an electron is emitted, it is accompanied by an antineutrino, for a combined lepton number of 0. When a positron is emitted, it is accompanied by a neutrino, again for a combined lepton number of 0.

When neutrinos and antineutrinos are taken into account, lepton number is conserved in all subatomic events that have so far been studied. Thus, the existence of the neutrino and antineutrino have not only rescued the three conservation laws of energy, momentum, and angular momentum; it also made it possible to establish a conservation law of lepton number.

Physicists could hardly avoid accepting the existence of these particles.

8

DETECTION OF NEUTRINOS

THE ABSORPTION OF PHOTONS

So far, the neutrino seems rather like a photon in many respects. Like the photon, the neutrino is uncharged, massless, and travels constantly at the velocity of light. Both are spinning particles. To be sure, the photon has a spin of $+1$ or -1, whereas the neutrino has a spin of $+\frac{1}{2}$ or $-\frac{1}{2}$, but this is not a particularly striking difference.

To find a striking, and even astonishing, difference between the two, let's first understand that it is always possible to consider the reverse of an event. For instance, a man holding a ball can throw it southward, let us say. If that ball were to reverse its path and approach the man, the man, by reversing his movements, could throw up his hand and catch it, bringing it to a halt.

In the original case, you would have the sequence: 1) man holding ball, 2) man throwing ball, 3) ball moving south. In the reverse case, you would have the sequence: 1) ball moving north, 2) man catching ball, 3) man holding ball. The situation would be like that of a strip of movie film that was first run forward and then backward.

Suppose we translate this principle to the subatomic world. For instance, if the electron of an atom drops from an excited state to a less excited state it will emit a photon of visible light, and the wavelength of that light will depend on the energy difference between the two excited states. That same atom can absorb (or "catch") a photon of that precise wavelength and, in response, the electron will be lifted from the less excited state to the more excited one.

Each type of atom can emit photons of various specific wavelengths (depending on the energy content of its particular group of excited states) and, under appropriate conditions, absorb photons of exactly the same specific wavelengths.

And yet the difference between an event and its reverse is not just a matter of change in direction and sequence. Catching a ball is more difficult than throwing it. In throwing a ball, you begin with a motionless object and set it into motion. There's no problem there. Since the object is motionless at the start it waits your convenience. You can take all the time you want, positioning the ball against your fingers, aiming it carefully, and so on.

In catching a ball, however, you begin with a moving object and now you cannot take all the time you want. You must snatch at the ball as it approaches; it will remain within your reach for only a portion of a second. In that portion of a second, you must put your hand accurately in the way of the ball and stop it. If you miss and the ball goes past you, you have not caught it.

It is the same for the atom emitting a photon. An atom which is about to emit a photon does so in a period of time that averages about 0.00000001 or 10^{-8} seconds. The atom, however, can choose its time, so to speak, and emit the photon at its convenience.

To absorb that same photon, the atom requires 10^{-8} seconds, this being the natural consequence of the reversibility of events. If it takes 10^{-8} seconds to emit a photon, it takes 10^{-8} seconds to absorb it.

But the atom can't absorb a photon without considerable trouble. The photon is moving at the speed of light and doesn't remain within reach of the atom for the entire time lapse of 10^{-8} seconds. In that period of time, a photon of light will travel 300 centimeters. You might picture the atoms it passes in that distance "snatching" at it and, generally, failing to catch it, much as baseball players would fail to catch balls hurtling by too quickly.

However, by sheer chance, one atom might catch and absorb the photon and if that happened, on the average, in 10^{-8} seconds, it would happen, again on the average, after a photon had traveled for 300 centimeters through matter. Some photons might travel a longer distance and some a shorter, but 300 centimeters would be the average.

But this supposes that the photon has no size of its own. Actually, it has considerable size. A typical photon of visible light has a wavelength of about 1/20,000 of a centimeter and this is about as wide as a thousand atoms laid side by side.

We can picture the photon of visible light, therefore, as a fuzzy sphere spread out over a thousand times the width of an atom and therefore possessing $1000 \times 1000 \times 1000$ or 1,000,000,000 times the volume of an atom. At any instant of time, the photon of light is in contact with about a billion atoms, any one of which can manage to catch and absorb it. The distance the photon can penetrate matter before being absorbed, therefore, is not 300 centimeters, but that distance divided by a billion; that is 0.0000003 or 3×10^{-7} centimeters.

Such a distance is not more than 10 or 15 atoms laid side by side. This means that a photon of light will not penetrate more than 10 to 15 atom depths into a solid before it is absorbed. Since 10 to 15 atom depths is nothing at all in the everyday world, most solids are opaque to light even in thin slices (though gold foil can be beaten so extremely thin as to become translucent to light).

The shorter the wavelength of light, the smaller the photon, the fewer the number of atoms it can be in contact with at any one instant, and the farther, therefore, it may travel through matter before being absorbed. It is for this reason that ultraviolet light can penetrate more deeply into human skin than visible light can; that X rays can pass all the way through the soft tissue of the body and are stopped only by the denser matter of bones; that gamma rays can penetrate through many centimeters of the densest matter.

(To be sure, visible light can pass through considerable distances of substances such as glass and quartz, to say nothing of most liquids, but this involves complications that need not concern us.)

THE ABSORPTION OF NEUTRINOS

We can now apply all this to the neutrino and antineutrino. The neutron in breaking down (to repeat once more) forms a proton, electron, and antineutrino. Let us write the equation again in order to have it clearly before us:

$$n \longrightarrow p^+ + e^- + \bar{v}$$

If we accept the thesis that a process can always be reversed under appropriate circumstances, we can say that it is possible for a proton to accept an electron and an antineutrino and become a neutron again. The reverse reaction would look like this:

$$p^+ + e^- + \bar{v} \longrightarrow n$$

Of course, we are now asking the proton to catch an electron and an antineutrino simultaneously, and this enormously increases the unlikelihood of a successful completion of the process. (It would be like asking an outfielder to catch two balls speeding toward him from different directions, and to catch them simultaneously with one hand.)

In order to remove some of the difficulty, let's add another turn to the reversal. Any process which involves the addition of an electron can proceed just as well if the subtraction of a positron is substituted. (This is analogous to the situation in algebra where adding $+1$ is the same as substracting -1.)

In other words, instead of expecting the proton to absorb both an electron and an antineutrino, we will expect it to absorb an antineutrino and emit a positron. The equation becomes:

$$p^+ + \bar{v} \longrightarrow n + \overline{e^+}$$

In this new version of the reaction, the conservation laws hold. In particular you can see that since a proton is replaced by a neutron (each with a baryon number of +1), and an antineutrino is replaced by a positron (each with a lepton number of −1, the laws of conservation of baryon number and of lepton number are conserved.

Now we need only consider the likelihood of a proton absorbing an antineutrino.

Remember that when a neutron breaks down to a proton, electron and antineutrino, the process has a half-life of 12.8 minutes. An individual neutron can take more than 12.8 minutes to break down or less than 12.8 minutes, but over a large number of neutrons, 12.8 minutes is the average time for breakdown.

It follows that for a proton to build back to a neutron by absorbing an antineutrino and emitting a positron, the same average time of 12.8 minutes is required. To put it more simply, it takes 12.8 minutes (on the average) for a proton to absorb an antineutrino.

But the antineutrino is traveling at the velocity of light and in 12.8 minutes, will travel 23,000,000,000,-000, or 2.3×10^{13} centimeters. This is equal to 140,-000,000 miles, which is about the distance from the Sun to Mars.

You might picture an antineutrino traveling through 140,000,000 miles of solid matter and being snatched at futilely all the time before it is stopped. Nor could things be made better by supposing the antineutrino to be spread out over a comparatively large volume as is true of the photon. The antineutrino is quite small; far smaller than a single atom.

In fact, matters are far more complicated than have

so far been indicated. In the case of photons, absorption is carried through by the electrons which make up most of the volume of the atom, and in solid matter, atoms are lined up solidly.

Antineutrinos, however, are absorbed by the protons of the atom and these are located only in the atomic nucleus, which makes up only the tiniest portion of the atomic volume. An antineutrino, streaking through solid matter, rarely approaches the tiny nucleus, so that it is close enough to a proton to be captured only about a hundred-millionth of the time it is within the atom.

For the antineutrino to have a proper chance of being absorbed, then, it must travel through a hundred million times 140,000,000 miles of solid matter. It has been estimated, in fact, that the average antineutrino will travel through about 3500 light-years of solid lead before being absorbed.

Naturally, one will not find a 3500-light-year thickness of solid lead anywhere in the universe. The universe is composed of individual stars, thinly spread out, and each far less than a millionth of a light-year in diameter. Most of them are composed of matter that is considerably less dense than lead, except for a relatively small core of super-dense material. A few stars are super-dense throughout but these are particularly small —no larger than planets. And even the super-dense portions of stars are not very efficient at stopping antineutrinos.

An antineutrino traveling through the universe in any direction will very rarely pass through a star and still more rarely through its super-dense core. The total thickness of stars it will pass through in a journey from end to end of the visible universe will be far less than a single light-year.

What I am saying about antineutrinos here applies, naturally, to neutrinos as well.

We can say then that neutrinos and antineutrinos are simply not absorbed in any significant fashion. Once formed in any subatomic process they merely travel on forever, unchanged and unaffected by their surroundings. Occasionally one may be absorbed, but those that are absorbed are so few in number as to be insignificant in comparison with the vast number that exist and that are continually being formed.

On the basis of present knowledge, it seems safe to say that virtually all the neutrinos and antineutrinos formed during the lifetime of the universe are still in existence today.

TRAPPING THE ANTINEUTRINO

This is all depressing news for the physicist. However much he may deduce the necessity for the existence of the neutrino and antineutrino from the conservation laws, he would nevertheless feel truly happy only if he could actually detect the tiny particles and demonstrate their existence by direct observation.

But to demonstrate their existence, he must first catch one; that is, he must cause it to interact with some other particle in such a way as to enable him to detect the results. Since it is virtually impossible to catch a neutrino or antineutrino, their existence, it would seem, cannot be demonstrated.

The result is that the physicist saves his concept of the structure of the universe, as he has developed it over three centuries, by insisting on the existence of something which he must accept on faith. He deduces the existence of the neutrino from the requirements of his theories, and he saves his theories by insisting on the existence of the neutrino. This is called "arguing in a circle."

As long as physicists were forced to maintain that position, they would remain particularly subject to doubt and uncertainty. It was terribly important, therefore, to work out some method for detecting the neutrino or antineutrino, if it could possibly be done.

The chink in the almost impenetrable armor of the elusive neutrino lies in the word "average." A few pages back I said that the average antineutrino will travel through 3500 light-years of solid lead before being absorbed. But that's the *average* antineutrino. Some antineutrinos would travel a shorter distance before being absorbed, and some a longer distance. A very few will travel a very short distance before being absorbed, and a very few will travel an unimaginably long one.

It is necessary to concentrate, then, on that infinitesimally small percentage of antineutrinos that will be absorbed in a thickness of matter that can be handled in the laboratory—say a few feet. In order to make that infinitesimally small percentage include as many antineutrinos as possible, the richest convenient source of those particles must be dealt with.

A particularly rich source is a nuclear fission reactor. In nuclear fission, a massive atom, such as that of uranium-235, breaks into two nearly equal parts. Smaller atoms require fewer numbers of neutrons in the nucleus than larger ones do; therefore the process of fission leaves some neutrons in excess. These neutrons, sooner or later, break down to form protons and electrons and antineutrinos. While the reactor is going at full speed a tremendous number of antineutrinos are continually being produced.

In 1953, experiments aiming toward the detection of the antineutrino were begun by a team of American physicists, led by Clyde L. Cowan, Jr., and Frederick Reines. For their source of particles, they made use of a nuclear fission reactor at Savannah River, South Carolina. This reactor gives off an estimated 1,000,-000,000,000,000,000, or 10^{18} antineutrinos each second.

The next step was to supply these myriads of antineutrinos with as generous a target as possible. The simplest way to crowd a great many single protons (the natural target) together is to use a quantity of water. Each water molecule contains two hydrogen atoms and an oxygen atom and the nuclei of the hydrogen atoms are single protons.

Cowan and Reines made use of five tanks of water

6¼ feet long by 4½ feet wide, but varying in thickness (see Figure 11). Two of the tanks were thin "target tanks" only three inches high. The other three were thick "detector tanks" two feet high. They were arranged in a "two-decker sandwich": detector/target/detector/target/detector. The water in the target tanks

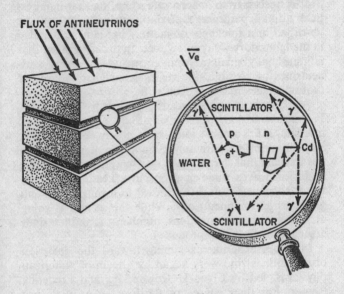

FIGURE 11. Detection of the Antineutrino

contained a small quantity of a dissolved chemical, cadmium chloride. The detector tanks contained a solution of a scintillator substance: that is, one which, on absorbing a subatomic particle, will gain energy and re-emit it almost at once as a flash of light.

This large two-decker sandwich of tanks was placed in the path of the antineutrinos pouring out of the fission reactor. Then it was only necessary to wait. If the antineutrinos actually existed, every twenty minutes (on the average) one of them ought to be absorbed by a proton.

But what then? The difficulty was to recognize that this very rare event had taken place, in the midst of all the other subatomic events going on within the water tanks at the same time. These tanks were bombarded continually by particles and radiation of all sorts arising from traces of radioactive substances in the air, the building materials, the soil all about, as well as from cosmic radiation invading Earth from outer space.

At first, the unwanted subatomic "noise" obscured possible detection of any antineutrino absorption. Gradually more and more efficient shielding was added, to keep out all unwanted radiation and particles. No amount of shielding, no thickness of metal or concrete, would keep out antineutrinos, of course, and eventually the "noise" had been reduced to a level that would not seriously mask the tiny whisper of the very occasional antineutrino that was captured.

But that whisper still had to be recognized. Consider that when an antineutrino is absorbed by a proton, there are produced a neutron and a positron—a distinctive combination of particles.

Once the positron is formed in one of the target tanks, it interacts with an electron in less than a millionth of a second and, in the process, produces two gamma-ray photons, each 0.51 Mev in energy content. The two photons must, by the law of conservation of momentum, be fired off in exactly opposite directions. If one of them leaves the target tank to penetrate the detector tank above, the other other must penetrate the detector tank below. In each detector tank a scintillation is produced. These scintillations, once produced, are automatically recorded by the hundred or more phototubes which are placed all round the water tanks.

And what about the neutron? Ordinarily, it might simply move about randomly, bouncing off water molecules (made up of oxygen and hydrogen atoms which have very little tendency to absorb a neutron) until it breaks down of its own accord, an average of 12.8 minutes after formation. It is, however, impractical to wait that long, especially since that time is only an average and the event could easily take place a few minutes sooner or later.

That is where the cadmium chloride in the target tank comes in. The neutron, after formation, goes bouncing about in a random path until it strikes a cadmium atom. It is then absorbed almost at once. This takes place several millionths of a second after the annihilation of the positron, a conveniently short lapse of time, yet long enough to allow a clear gap between the positron event and the neutron event.

When the neutron is absorbed by the cadmium atom, energy is gained which is almost immediately emitted in the form of three or four photons with energies that total 9 Mev.

That is the pattern, then, that Cowan and Reines were watching for: the simultaneous appearance of two photons of 0.5 Mev each, activating two phototubes at opposite sides of the water tanks, followed within a few millionths of a second by the simultaneous appearance of three photons of 3 Mev each (sometimes four photons of 2¼ Mev each).

No other single subatomic interaction would (as far as was known) give rise to exactly this combination of events. If this combination were detected, it would be at least reasonable to conclude that an antineutrino had been absorbed by a proton and, therefore, that antineutrinos actually existed.

Such a combination of events was indeed detected, but then another thought arose in the cautious minds of the experimenters. What if such a combination of events were produced not by one subatomic interaction but two? Suppose a positron were produced in one way or another and that a few millionths of a second later a cadmium atom happened to absorb a neutron that was in existence without any connection with the positron at all. In that case, the two photons followed by three photons would not involve a single interaction (which could only be that of an antineutrino with a proton) but two unconnected ones (which could be anything). How could one tell?

The experimenters settled that matter by running their tests first with the nuclear fission reactor operating and then with it shut down. If the fission reactor were shut down, the tanks would still be subjected to "noise." The only thing that would stop would be the antineutrino bombardment. (Actually, there are antineutrinos in the general environment at all times, but in quantities much smaller than those in the immediate neighborhood of an operating fission reactor.) With the reactor shut down, therefore, the two-event coincidences would continue, but the single-event antineutrino absorption would stop.

It turned out that there were 70 fewer detections of the proper pattern per day with the reactor shut down than with the reactor running. That means 70 antineutrinos were absorbed and detected per day (one every twenty minutes). The matter had moved into the region which could be described as "demonstrated beyond a reasonable doubt," and in 1956 the announcement was made that the antineutrino had finally been detected, fully twenty-five years after Pauli had first predicted the existence of such a particle.

This feat is usually spoken of as the "detection of the neutrino" although it was the antineutrino that was detected. However, with the antineutrino detected, physicists took it for granted that the actual existence of the neutrino could be considered as having been demonstrated also.

9

NEUTRINO ASTRONOMY

ANTINEUTRINOS AND THE EARTH

Once the neutrino's existence was demonstrated, it became more reasonable to think of the relationship of the neutrino to the universe as a whole. In other words, a new branch of science, *neutrino astronomy,* came into being.

One great natural source of neutrinos in the universe is the radioactive atoms. Radioactive transformations, occurring among isotopes that are unstable but with such long half-lives that they have existed all through the Earth's lifetime, almost always involve the production of beta particles, or of alpha particles followed by beta particles. The reverse case, the emission of positrons, does not take place among the radioactive atoms that occur naturally in the Earth's crust.

This means that natural radioactivity produces antineutrinos only, one antineutrino accompanying each beta particle.

The natural production of beta particles on Earth comes about through the breakdown chiefly of three particular types of atoms: uranium-238, thorium-232, and potassium-40. The uranium-238 breaks down in stages to lead-206, giving off, in the process, eight alpha particles and six beta particles. The thorium-232 atom breaks down, in stages, to lead-208, emitting seven alpha particles and four beta particles. The potassium-40 atom breaks down to a calcium-40 atom, giving off a single beta particle.

Allowing for the difference in mass of these atoms, for their half-life, and for their percentage occurrence in the Earth's crust, it has been estimated that for each kilogram of that crust, 7330 beta particles are produced per second, of which 6200 can be attributed to uranium-238, 800 to potassium-40, and 330 to thorium-232. The contribution of all other radioactive atoms (including the rather rare uranium-235) are so small in comparison that they may be ignored.

For each beta particle, an antineutrino is emitted, so we can say that the Earth is continually witnessing the emission of 7330 antineutrinos for each kilogram of its crust. On that basis, the Earth's crust as a whole is emitting about 175,000,000,000,000,000,000,000,000,000 or 1.75×10^{26} antineutrinos each second.

This includes the crust only, of course. Uranium, thorium, and potassium are highly concentrated in the Earth's surface layers, but there must also be some through the vast deeper layers of the planet and these must certainly contribute to the antineutrino production. However, estimates as to the content of these elements in the deeper layers are, at best, very rough, and little can be done with them.

Even if we consider only the 1.75×10^{26} antineutrinos produced each second by the Earth's crust, the results are impressive. Half of them, more or less, must be emitted in a generally downward direction and must then pass through the Earth. If these are imagined to be distributed evenly through the body of the Earth, then it would seem that any portion of the Earth 40 cubic centimeters in volume (about 2½ cubic inches) ought to contain one antineutrino during any given one-second interval of time. Since the average human body contains about 70,000 cubic centimeters, it follows that each second some 1750 antineutrinos are passing through our bodies.

Naturally we are not aware of it, since the antineutrinos do not interact with the atoms within our body. During a seventy-year lifetime, the chances are only one in a billion or so that even a single antineutrino would be absorbed by one of the protons of a human body.

In fact, the antineutrinos pass through the entire body of the Earth as easily as they pass through our bodies and if a handful of them are absorbed in the process, the number is completely insignificant compared to all those that get through.

NEUTRINOS AND THE SUN

Next let us consider the fusion reactions that go on in the interior of stars. In a star like our Sun, for instance, energy is obtained out of the conversion of hydrogen to helium. The details of this conversion can vary but the overall reaction involves the conversion of 4 hydrogen nuclei (each with a charge of $+1$ for a total charge of $+4$) into a helium nucleus (with a charge of $+2$).

In order to conserve electric charge, two positrons are also formed, carrying a charge of $+1$ each. In this way, we begin and also end with a charge of $+4$.

The four hydrogen nuclei with which we begin have a lepton number of zero, as does the helium nucleus with which we end. The two positrons have a lepton number of -1 each, however, and if a lepton number is to be conserved, they must be balanced by the simultaneous formation of two neutrinos with a lepton number of $+1$ each. We can conclude, then, that for every two atoms of hydrogen consumed in the Sun, one neutrino (*not* an antineutrino) is produced.

The total number of neutrinos produced by the Sun depends on the total number of hydrogen nuclei consumed. In the fusion of hydrogen to helium, 0.71 percent of the mass is converted to energy and, as I have already said, (see page 57) the Sun is losing 4,600,000 tons of mass each second. If this loss represents 0.71 percent of the total mass of hydrogen being fused to helium each second then that total mass of fused hydrogen is 650,000,000 tons. One can calculate the number of hydrogen nuclei in 650,000,000 tons of the substance and we can then conclude that every second 360,000,000,000,000,000,000,000,000,000,000,000,000,000 or 3.6×10^{38} hydrogen nuclei are being fused in the Sun and half that many or 180,000,000,000,000,000,000,000,000,000,000,000,000, or 1.8×10^{38} neutrinos are being produced.

The number of neutrinos being produced by the Sun each second is, therefore, just about a trillion times greater than the number of antineutrinos being produced by the Earth's crust each second. This is probably a pretty fair weighting of the comparative productions of neutrinos and antineutrinos by stellar fusion and by radioactivity in general. On the whole, though the universe is riddled by a constant streaking of both neutrinos and antineutrinos, the number of neutrinos far, far outweighs the number of antineutrinos (assuming the universe to consist of matter only). Once again, then, our universe witnesses a condition in which a particle is very common and the corresponding antiparticle is (by comparison) quite rare.

In a universe made up of antimatter, natural radioactivity would produce neutrinos and stellar fusion would produce antineutrinos. There it would be the antineutrinos that would be by far the more common.

From this standpoint, we can understand that the energy of the Sun is poured out into the universe about it very largely in the form of both photons and neutrinos, and not photons alone. To be sure, most of the energy is carried off by the photons. Perhaps 5 percent, no more, is carried off by the neutrinos.

Yet there is an interesting difference in the manner in which photons and neutrinos carry off that energy. The photons produced in the incredibly hot center of the Sun, where the fusion processes take place at a temperature of approximately 15,000,000° C, are easily absorbed by the surrounding material. They only travel a centimeter or so before being reabsorbed. They are then re-emitted of course, but are then again reabsorbed, and so on.

Only very slowly do photons make their way outward from the center, through over 400,000 miles of solar matter, to the surface. The solar matter is, indeed, an excellent heat insulator for that reason, so that the Sun's surface is merely 6000° C. This is hot enough by earthly standards, to be sure, but consider that this 6000° C surface is only a little over 400,000 miles from a volume of matter at a temperature of 15,000,-000° C.

Neutrinos do not have to face the tortuous travels of the photons. At the instant each is formed it shoots away at the velocity of light, oblivious to the matter of the Sun and unabsorbed by it except in an exceedingly small percentage of cases.

In no matter which direction the neutrino darts away from the Sun's center, it will be at the surface within three seconds. It will then shoot out into space and (if it is aimed in the right direction) will reach the Earth in eight minutes. It will pass clear through the Earth in $\frac{1}{25}$ of a second at most, then move on farther in an endless journey.

The solar neutrinos travel in every direction, of course, and all but a tiny fraction miss the Earth, which offers a small target indeed at its great distance from the Sun. Even so, the Earth receives a healthy 80,000,000,000,000,000,000,000,000,000,000 or 8×10^{28} neutrinos every second from the Sun. This is some 450 times as great a number of neutrinos as is the number of antineutrinos being produced by the radioactivity of its own crust. Each square centimeter of the Earth's cross-sectional area receives about 60,000,-000,000 or 6×10^{10} neutrinos each second.

This activity provides the major portion of the neutrinos we receive. Others are received from every star in the sky, but the stars are so much farther away than the Sun that their neutrinos are spread very thinly by the time they reach us; exceedingly few intersect the Earth itself. (To put it another way, the Earth is a much smaller target as viewed from Alpha Centauri than from the Sun; and from more distant stars it is a tinier target still.)

TRAPPING THE NEUTRINO

Since the Sun is so rich a source of neutrinos, it would be very satisfying if these could be detected in some way analogous to that in which antineutrinos were detected.

The antineutrino, remember, is absorbed by a proton, yielding a neutron and a positron (see page 113). To detect the neutrino we must set up a kind of mirror-image reaction. That is we must have the neutrino absorbed by a neutron to form a proton and an electron:

$$v + n \longrightarrow p^+ + e^-$$

In the case of the antineutrino we had to collect a target rich in individual protons, and in this case we would have to collect one rich in individual neutrons. Unfortunately, whereas individual protons are easily collected in the form of hydrogen or of a hydrogen-containing chemical compound such as water, individual neutrons cannot be collected in large quantities.

Instead, use is made of neutron-rich atomic nuclei. One possibility (following a suggestion of the Italian physicist Bruno Pontecorvo) is chlorine-37, which makes up about ¼ of all chlorine atoms. Its nucleus contains 17 protons and 20 neutrons. If one of those neutrons absorbs a neutrino, it becomes a proton (and emits an electron). The nucleus will then have 18 protons and 19 neutrons and will be argon-37.

To form a sizable target of chlorine-neutrons, one might use chlorine gas itself, or, better, liquid chlorine, since in the liquid form more chlorine molecules (each molecule made up of two chlorine atoms) can be packed into a given volume. However, chlorine is a very corrosive and toxic gas, and to keep it liquid would present a problem in refrigeration.

Instead, chlorine-containing organic compounds can be used. (There is no reason why nuclear reactions generally cannot proceed with atoms existing in molecules as well as they can with free atoms.) Carbon tetrachloride, with molecules made up of one carbon atom and four chlorine atoms, can be used; so can perchloroethylene, with molecules made up of two carbon atoms and four chlorine atoms. They are liquids at ordinary temperatures and are quite safe to handle with ordinary precautions. (In fact, they are common dry-cleaning fluids.)

If a chlorine atom, forming part of the perchloroethylene molecule, absorbs a neutrino and becomes an argon atom, it must release its hold on the molecule, for argon atoms will not combine with any others. Thus, neutrino absorption will form free argon atoms out of bound-to-a-molecule chlorine atoms. Free argon atoms will eventually collect as tiny bubbles of gas.

Not many argon atoms would be formed, of course, and it is natural to wonder how their appearance would be detected. The detection possibility arises from radioactivity. Chlorine-37 is a perfectly stable atom, but argon-37 is not. Argon-37 is radioactive and its presence can therefore be detected even in small concentration and identified through the particular form of its radioactivity.

To give argon-37 the best possible conditions for exhibiting its radioactive properties, however, it should be concentrated as much as possible. One way of doing this is to wait a few weeks to allow as much argon-37 as possible to accumulate, then flush the tanks with helium gas. The helium (a gas that is quite similar to argon) will sweep out the argon-37 atoms with it, and those atoms will be easier to detect when concentrated in the helium than when spread out through the original perchloroethylene.

The American physicist Raymond R. Davis has used the chlorine/argon technique to demonstrate that there really are two neutrinos—the neutrino itself and the antineutrino, as is required by the law of conservation of lepton number (see page 107).

Suppose there weren't two such particles, but only one, and that the neutrino was its own antiparticle, as the photon is. If this were so then the same particle would be both neutrino and antineutrino and would be expected to do, with equal ease, the tasks we now attribute, separately, to each. For instance, if we find a particle which is absorbed by protons, yielding positrons and neutrons, that same particle also ought to be absorbed by chlorine atoms and form argon atoms. The former is an antineutrino property and the latter a neutrino property, and if a single particle is both, it should perform both functions.

A nuclear fission reactor gives off particles that are absorbed by protons in the proper manner and that are therefore antineutrinos. Would the same particles also convert chlorine atoms to argon atoms? Davis set up tanks of perchloroethylene near a fission reactor in 1956 and did not detect any such reaction. The antineutrinos definitely proved to be present by Cowan and Reines could not, it seemed, fulfill a neutrino function; there had to be a separate particle which served as a neutrino. The law of conservation of lepton number was upheld.

The next step is to detect the solar neutrinos directly. For this purpose, *neutrino telescopes* have been prepared. These consist of huge tanks of 100,000 or more gallons of perchloroethylene which are being placed far underground. Davis is working with one a mile deep in a silver mine. In 1965 the detection of neutrinos from outer space was reported by Reines, working in a South African gold mine. Seven neutrinos were detected in nine months.

It may seem odd at first that astronomic observations are to be conducted deep in the bowels of the Earth, but it makes perfect sense. Very little in the way of subatomic "noise" can penetrate a mile of the Earth's crust. Even cosmic-ray effects are shielded off and all that is left is the effect of the trace radioactivity in the mine surroundings themselves. Solar neutrinos on the other hand can reach the tank of perchloroethylene without trouble even though it is a mile deep. They would reach it with equal ease if it were at the Earth's center.

The value of detecting solar neutrinos would be great. Photons from the Sun do not reach us directly from the solar center but only by way of incredibly involved travels through the matter of the Sun, travels that introduce vast changes in the photon properties. Neutrinos, on the other hand, reach us directly from the center.

From the energies of the neutrinos received, it is hoped, physicists will be able to deduce, in full detail, the nature of the fusion reactions going on there. The actual energies of the neutrinos formed depend on the exact route taken in the passage from hydrogen to helium. From the energy spectrum of the neutrinos observed, the route can be deduced, and from that the internal temperature of the Sun can be measured, and perhaps many other characteristics as well.

In short, we will be able to "see" directly into the Sun's center and, it seems fair to hope, learn a great deal concerning matters now obscure.

SUPERNOVAS AND NEUTRINOS

Neutrino astronomy might well yield useful results outside the solar system, too.

In the last few decades, astronomers have worked out, in considerable detail, theories as to the nuclear changes that go on in the cores of the stars that are past their youth. Our own Sun, one must note, is not one of these. Although it is five or six billion years old, it is still in its youth; it is still fusing its ample hydrogen content to helium. Such hydrogen-fusing stars are very stable and can continue for many billions of years with little change.

As hydrogen fusion continues, however, the core of helium grows at the star's center, increasing both in volume and in temperature. Eventually as the temperature increases, nuclear processes which are of insignificant importance in the Sun become crucial. At

15,000,000°, for instance, the temperature of the Sun's center, helium atoms do not themselves undergo fusion at a rate more than microscopically slow. When the temperature reaches 100,000,000°, however, three helium nuclei begin to combine to form a carbon nucleus at a respectable rate. The star progresses to the stage of "helium-fusing" (see Figure 12A).

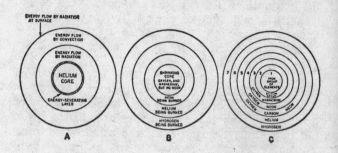

FIGURE 12. Stages of Stellar Evolution

A new carbon core forms within the helium core and its central temperature continues to increase. At 600,000,000° carbon atoms begin to fuse to magnesium. At 2,000,000,000° magnesium atoms fuse to sulfur, and at 4,000,000,000° sulfur atoms begin to fuse to iron. (At every step there are also a number of complicated side-reactions.)

When a star begins to burn helium and to form successively more complicated nuclei, it is in its last stages (see Figure 12B). The fusion of nuclei from helium on produces comparatively little energy. In the fusion of hydrogen to helium, the energy produced for each helium nucleus formed is 27.5 Mev. In passing from helium to iron, however, less than 9 Mev is formed per helium nucleus consumed.

This means that when a star has passed beyond the hydrogen-fusing stage, it has used up about ⅔ of all its easily available nuclear energy. The later changes must proceed all the faster to produce enough energy to keep the star going, particularly at the higher and higher temperatures. A star may pass from the magnesium-fusing stage to the sulfur-fusing stage in a mere century or less.

And the sulfur-fusing stage is the last (see Figure 12C). When iron is formed, a star has reached a dead end as far as nuclear reactions are concerned, since the iron nucleus is as compact as a nucleus can be and contains a minimum of energy. To convert an iron nucleus into any other kind, whether by fusion to more complicated nuclei, or by fission to less complicated ones, requires an input of energy.

Astronomers speculate that the only place from which such an input of energy can arise is the gravitational field. Once again the old Helmholtz theory of gravitational contraction as a source of radiation arises, but in a vastly changed and more catastrophic form. To supply the energy required to change the iron back into helium, gravitational collapse must proceed at a phenomenal rate. The star must collapse to a tiny fraction of its previous volume in a day or less.

As a result of this sudden and drastic collapse, the material of the star condenses to the point where nucleons are thrust together much more closely than in ordinary matter. The star becomes a *white dwarf* made up of *degenerate matter* so dense as to have a mass of hundreds of tons per cubic centimeter.

In the process of this collapse, the material in the outer reaches of the star is also compressed drastically. In those outer regions, there still remain scraps of the original fuel of the star, even hydrogen. As that fuel is compressed, its temperature zooms and it undergoes fusion with one vast thunderclap, so to speak.

The result is a *supernova,* a star which, for a brief time, radiates energy at a rate many millions of times that of an ordinary star. For a few weeks it may radiate energy at a rate equivalent to that of an entire galaxy of ordinary stars.

The vast catastrophe of a supernova is, perhaps, essential to the structure of the universe. Astronomers currently believe that the universe in its youth consisted only of hydrogen. The more complicated atoms, up through iron, were then slowly formed in the cores of stars. Each supernova, thanks to the vast energies suddenly made available, could form all the nuclei more complicated than that of iron and then, in exploding, spread all the atoms it had formed through space.

As a result, thanks to all the supernovas that have exploded since the universe was first formed, the thin matter spread between the stars is comparatively rich in heavy atoms. Stars that slowly form out of this interstellar matter (that had once been part of earlier stars) are *second-generation stars*. Such second-generation stars are comparatively rich in heavy atoms even though the processes going on in their cores do not form them. The Sun is such a second-generation star, and the Earth exists in its present form only because the atoms composing it were once formed by stars that have long since exploded and which now exist, if at all, as white dwarfs lost in the vastness of space.

Astronomers try to deduce, from the relative concentration of the various types of nuclei in the universe, just what reactions may proceed in the center of stars at various stages of their history. That serves as one of the foundations for the speculations (still highly tentative) concerning the evolution of stars.

Supernovas do not occur frequently. It is estimated that in any given galaxy, three will occur per thousand years. It is thought that three supernovas have occurred in our own Milky Way Galaxy in the last thousand years, the most recent occurring in 1604 just before the telescope was invented. We are "due" for one anytime now and we can only hope that it will hold off until the science of neutrino astronomy is further advanced.

Hope that the necessary progress will not require much more time arises out of the work of the American physicists Phillip Morrison and Hong-Yee Chiu at Princeton. In searching for new nuclear interactions that might help explain the course of stellar evolution, they began to consider the nature of electron-positron interaction.

The ordinary result of such an interaction is the formation of two gamma-ray photons (see page 86). However, there was just a small chance that an electron and a positron might interact to form a neutrino and an antineutrino instead. The interaction could then be symbolized thus:

$$e^- + \overline{e^+} \longrightarrow \nu + \overline{\nu}$$

All the conservation laws would be conserved in such a reaction. For instance, electric charge is conserved since $-1+1=0+0$. And lepton number is conserved, too, for $1-1=1-1$.

Under ordinary conditions, the formation of neutrinos from electron-positron interaction is extremely unlikely. For every neutrino/antineutrino pair formed in this fashion, at least 100,000,000,000,000,000,000 or 10^{20} photon pairs are formed. It might seem, therefore, that neutrino formation in this manner can be ignored.

This is true, but only under ordinary conditions. The center of a star is not an example of an ordinary condition; and as the center of a star grows hotter and hotter in the course of its evolution, it becomes less and less ordinary.

Morrison and Chiu calculated, from theoretical considerations, that neutrino/antineutrino pairs were formed in greater and greater proportions from electron-positron interactions, as temperature increased. Furthermore, whereas the photons were retained in the center of the star for long periods of time, the neutrino/antineutrino pairs escaped in seconds.

The neutrino/antineutrino pairs represented an "energy-leak" in the star, and as the star grew older, it grew leakier. By the time a central temperature of 600,000,000° was reached, fully half the star's energy was being radiated in the form of neutrinos and antineutrinos. At temperatures above 600,000,000°, a star might fairly be called a *neutrino star*.

Such a star is close to the ultimate supernova collapse, and the radiation of neutrinos actually should contribute to this collapse. By the time the ultimate temperature of 6,000,000,000° is reached and an iron core capable of yielding no more nuclear energy is forming at the star's center, neutrinos are being formed in such numbers as to make the star catastrophically leaky. Chiu estimates that at that temperature, the escaping neutrinos can carry off all the internal store of energy of the star in a single day.

It is the star's internal store of energy that keeps it expanded against the compressing influence of its own gigantic gravitational field. Once the neutrinos carry away that internal store there is nothing for the star to do but collapse and so comes the supernova.

Under these conditions, the actual supernova stage is preceded for a short time, perhaps not more than a few centuries, by an unusually high emission of neutrinos. If the neutrino telescope works, it may turn out that some portions of the heavens are unusually good sources of neutrinos. (Chiu estimates that a source as much as 1000 light-years away might yield the capture of one neutrino every 10 to 100 seconds.) Those sources may be watched for the appearance of an imminent supernova (or, perhaps, for catastrophes of other types).

So far in the history of astronomy, supernovas have been studied only after they had exploded. If neutrino astronomy makes it possible to study supernovas immediately before the explosion, floods of information concerning stellar interiors might be revealed. The excitement of astronomers would be indescribable.

THE UNIVERSE AND NEUTRINOS

Indeed, neutrino astronomy may even tell us more general facts about the universe.

As I have said, in a universe of ordinary matter

neutrinos predominate and in one of antimatter, antineutrinos would predominate (see page 125).

What, then, can we say about the neutrino population of the universe generally? As yet, nothing positive, but we can speculate. Since all stars form neutrinos (or antineutrinos if the stars are made of antimatter) and since these are virtually never absorbed, the universe contains huge floods of neutrinos moving through every part. These floods represent events that have taken place through all of the universe's history and that are still being added to.

This "neutrino background" of the universe might well be its dominating feature, if it could only be detected. If we had the faculty of seeing neutrinos, we might find the universe consists of virtually nothing but neutrinos, and that all other particles are minor impurities, so to speak, against this background. It may be only because we can't sense the background, that the impurities seem to be the universe.

Yet once we begin to detect occasional neutrinos we might be able to judge a great deal about the universe and its history that now we can only guess at or, perhaps, do not even conceive.

One obvious piece of strategy would be to measure the relative numbers of neutrinos and antineutrinos reaching us from the universe generally. If we are receiving neutrinos only, then we are living in a universe consisting of ordinary matter only, with an insignificant admixture of antimatter. If we insist on thinking that antimatter must exist in equivalent quantities with matter (because of the laws of conservation of baryon number and of lepton number) then that antimatter may be indetectable and part of a separate "antiuniverse."

If, instead, we are being bathed in floods that are predominantly antineutrino in nature, then the universe, generally, is composed of antimatter. The Milky Way Galaxy in which our Sun exists would then happen to be one of the very few galaxies which might happen to be made up of ordinary matter.

Again, it may be that neutrinos and antineutrinos reach us in roughly equal numbers. In that case, the universe is composed of roughly equal amounts of matter and antimatter. In this case, we might even be able to detect individual matter-galaxies and antimatter-galaxies by narrowing down the points in the heavens from which neutrinos and antineutrinos originate.

But the great findings of neutrino astronomy are, as yet, in the future, and in the lack of something concrete, speculations can be almost limitlessly startling. A suggestion made in 1962 on the basis of legitimate mathematical reasoning, for instance, was to the effect that neutrinos might be capable of traveling into the past.*

But facts, as they turned up, outran speculation in some ways. In that same year of 1962, a discovery involving the neutrino was made which, while not quite as far-out as time-travel, was unexpected and startling enough. To explain how that came about, I must begin by taking another look at the atomic nucleus in some detail.

*This sounds amazingly like science-fiction. Indeed the plot of my very first science-fiction story, written in 1937 and never published, dealt with the supposed ability of subatomic particles to move through time.

10

THE NUCLEAR FIELD

REPULSION WITHIN THE NUCLEUS

In 1932 the atomic nucleus came to be looked on as made up of protons and neutrons exclusively (see page 74). Earlier theories that electrons were present in the nucleus were discarded.

Although this solved many problems, it also raised one that had not existed before. As long as physicists believed electrons were present in the nucleus, they had labored under the belief that they understood what kept the nucleus together. With the electron eliminated, that belief was eliminated as well.

Protons all carry positive charges and should therefore repel each other. The presence of electrons in the nucleus would have introduced an attractive force, since protons and electrons, carrying unlike charges, would attract each other. Electrons would then, conceivably, act as a "nuclear cement." Neutrons, on the other hand, being electrically uncharged, neither attracted nor repelled protons and could not, it seemed, serve as such a nuclear cement.

Nor was the repulsion between protons a weak force. In 1785 Coulomb (whose name later came to be given to a unit of charge, see page 66) had shown that the force of repulsion between two positively charged objects could be expressed by the following equation:

$$F = \frac{q_1 q_2}{d^2}$$

where q_1 and q_2 are the electric charge on the two objects in electrostatic units (see page 67) and d is the distance between them, center to center, in centimeters. When these units are used, F represents the force of repulsion in a unit called *dynes*.

In the case of the two protons, the electric charge on each is equal to 0.000000000480298 or 4.80298×10^{-10} electrostatic units. Within the nucleus two neighboring protons are virtually in contact and are therefore separated, center to center, by a distance of about 0.0000000000001 or 10^{-13} centimeters. If we insert these figures into Coulomb's equation it turns out that two protons within a nucleus repel each other with a force of about 24,000,000 or 2.4×10^7 dynes.

Is there any way of counteracting this enormous repulsion? In the early 1930s only two kinds of forces were known. One was the result of electromagnetic interactions (of which the repulsion of one proton by another is an example) and the other is the result of gravitational interactions.*

The gravitational interaction is always, as far as we know, one of attraction. This means that two protons attract each other gravitationally, in addition to repelling each other electromagnetically. Can this gravitational attraction counteract the electromagnetic repulsion?

Newton had expressed his law of universal gravitation in 1687 in much the same form as Coulomb was to express his law a century later. Newton devised this equation for two objects experiencing gravitational interaction:

*You might wonder where the ordinary "mechanical forces" such as that which you exert on a ball when you throw it, fit in. Actually, when the atoms in one object approach the atoms in another, the electrons in the outskirts of the atoms in one object repel the electrons in the outskirts of the atoms in the other object. In exerting an ordinary push or pull, you are making use of these repulsions so that mechanical forces are examples of electromagnetic interactions. The forces that hold atoms together within molecules, and molecules together within larger systems are also the result of electromagnetic interactions.

$$F = \frac{Gm_1m_2}{d^2}$$

where m_1 and m_2 are the masses of the two bodies, and d is the distance between them, center to center, in centimeters. As for G, that represents the *gravitational constant*.

Newton did not know the value of G. That was not determined until 1798 (seven decades after Newton's death) by the English chemist Henry Cavendish. The best value of G now available (in units that fit in with those used for the other quantities in the equation) is 0.00000006670 or 6.670×10^{-8} dyne-cm^2/gm^2.

The mass of each proton is, of course, extremely small, 0.00000000000000000000000167252 or 1.67252×10^{-24} grams. The distance between the two protons within the nucleus can still be considered 10^{-13} centimeters.

Substituting all these values in the right-hand side of Newton's equation, we can determine the value of F, which will give us the force of gravitational attraction between the two protons, in dynes. It turns out that this gravitational attraction amounts to the insignificantly tiny quantity of 0.0000000000000000000000000000 186 or 1.86×10^{-29} dynes.

In other words, the electromagnetic interaction tending to push the protons apart is over a hundred thousand trillion trillion times as strong as the gravitational interaction tending to pull them together.

This is an indication of the extent to which the electromagnetic interaction is stronger than the gravitational interaction. It is no wonder that in considering the behavior of subatomic particles, gravitational interactions are neglected.

The wonder might arise that we ourselves, in ordinary life, are so conscious of the gravitational effect.

The explanation is that there is only a gravitational attraction and no gravitational repulsion. In the case of electromagnetic interactions there are both, along with two kinds of electric charge, so that the attractive effect can be made to neutralize the repulsive effect on the large scale. Any sizable body usually has a net electric charge of just about zero. The Earth and the Sun, for instance, have net electric charges of zero and do not interact at all electromagnetically.

The gravitational interaction, however, producing only attractive effects, becomes more noticeable with the increase in size of the body. However weak the attraction may be, it piles up without neutralization as mass increases. For objects the size of planets and stars, gravitational attraction becomes tremendous. We are therefore always aware of the Earth's pull on us and mistakenly think of gravitation as a strong force whereas actually it is an incredibly weak one.

ATTRACTION WITHIN THE NUCLEUS

If we neglect gravitational interactions in considering the atomic nucleus and consider it in the light of electromagnetic interactions only, then the nucleus cannot exist. Its constituent particles could not be brought together in the first place against the tremendous repulsive forces among the protons; and if they were brought together somehow, they would immediately move apart in what could only be a tremendous explosion.

Only the hydrogen nucleus, consisting of a single proton (or, in some cases, of a proton plus a neutron), could exist under these conditions.

And yet all sorts of complex nuclei are formed and do exist and remain stable. The nucleus of uranium-238 contains 92 protons mashed together in virtual contact, yet it breaks down only with excessive slowness; while the lead nucleus, with 82 protons, is completely stable over all eternity, as nearly as we can tell.

If facts contradict theory, then theory must change. If protons hang together within a nucleus there must be an attraction holding them together; an attraction that more than counteracts the electromagnetic repulsion. A set of *nuclear interactions* giving rise to the necessary attraction must exist.

We can even predict what some of the properties of the nuclear interactions must be. In the first place, it must, as noted, be stronger than the electromagnetic interaction and must produce attractions between proton and proton (and for that matter between proton and neutron and between neutron and neutron). Secondly, the nuclear interactions must exert their influence over very short distances only. Let's see what this implies.

An electromagnetic interaction, or a gravitational one, can make itself felt over a considerable distance. It is as though each unit of electric charge serves as the center of an *electromagnetic field,* spreading out indefinitely in all directions and weakening with distance only gradually. Each unit of mass serves, similarly, as the center of a *gravitational field.*

The strength of each of these two fields decreases as the square of the distance between the interacting objects. If the distance between protons is increased 2 times, the gravitational attraction and the electromagnetic repulsion both decrease 4 times. If the distance increases 3 times, the field intensities decrease 9 times and so on.

This is a rapid fall-off but not a very rapid one. Despite the steady weakening, both fields can make their effects felt over long distances. This is best seen in the case of the gravitational field, for the Earth is held firmly in the grip of the Sun's gravity, despite the fact that 93,000,000 miles separates the two. For that matter the far more distant Pluto is also held in the Sun's grip, and the Sun is held in a huge orbit about the center of the Galaxy which is much, much farther away still.

Electromagnetic and gravitational fields may both, therefore, fairly be termed "long-range."

The nuclear interactions must originate in a *nuclear field* which however cannot exert its effects in the same "inverse-square" fashion. Two protons are mutually attracted with great force under the influence of the nuclear field as long as they are in virtual contact, but at distances greater than the width of the atomic nucleus, the attractive effect produced by the nuclear field is already weaker than the repulsive effect of the electromagnetic field; for, except inside the nucleus, two protons invariably exert only a net repulsion on each other.

Indeed, if an atomic nucleus is of unusually large size, the nuclear attraction does not extend comfortably over the entire width of the nucleus, which then shows a distinct tendency to break apart under the stress of electromagnetic repulsion. It is this which probably accounts for the ejection of small portions of the nucleus in the form of alpha particles on the part of the elements with unusually complex nuclei, and for their ability to undergo, on occasion, the even more radical breakup we call "fission."

The nuclear field may weaken not as the square of the distance but as the seventh power. If the distance between two protons is doubled, the attraction between them is decreased not 2^2, or 4 times, but 2^7 or 128 times. If this is so then within the nucleus the nuclear field can be over a hundred times as intense as the electromagnetic field, and outside the nucleus the nuclear field can be neglected.

One other point can be made. In 1932 Heisenberg (who first suggested the proton-neutron model of the nucleus) worked out a theory which made it seem that interactions produced by particular fields were brought about through the exchange of particles. In an electromagnetic field, for instance, attraction and repulsion came about through the exchange of photons between the bodies experiencing such attraction or repulsion. These are called *exchange forces*.

If Heisenberg's reasoning applied to the nuclear field as well, then some particle ought to be exchanged among the protons and neutrons of a nucleus in order to bring about the necessary attractions that held them together.

But what should this nuclear exchange particle be like and why should it produce a short-range force?

The answer (like so many other answers in nuclear physics) arose out of a consideration of the conservation laws again, but this time those conservation laws took on a startling new aspect.

THE UNCERTAINTY PRINCIPLE

My discussion so far of the conservation laws implies that they must hold exactly. We might deduce that this is so for we could argue that if energy, let us say, or momentum, were created or destroyed in even the tiniest amount, all sorts of phenomena would occur that are not, in actual fact, observed.

But suppose you were not content with the mere *deduction* that the conservation laws must hold exactly. Suppose, instead, that you tried to make actual measurements in order to demonstrate the fact.

In measuring any property of a system, however, you must involve yourself with that system. In so doing it is inevitable that you affect the system in some fashion; and this may, in turn, disturb the very measurement you are trying to make.

An example of this occurs in the measurement of the temperature of a cup of hot water. You can measure that temperature by inserting a thermometer. The thermometer warms up to the temperature of the water and you then read the height of the mercury thread. As the thermometer warms up, however, it extracts heat from the water, which cools down slightly. The temperature you measure is therefore not quite the temperature before you inserted the thermometer.

Another example is the measurement of the air pressure inside an automobile tire. A small gauge is inserted in the valve and the air inside pushes out the inner cylinder of the gauge. You then read the pressure from the degree to which the cylinder has been pushed out. However, some of the air from within the tire has escaped in the process of pushing out the cylinder, so that the pressure you measure is not quite the pressure that existed before the measurement.

This holds generally true for any form of measurement but, of course, in ordinary life the method of measurement can always be made fine enough to prevent any significant change in the property being measured.

Still, suppose you were not satisfied with just preventing "any significant change." Suppose you wanted no change at all in the property. Could you refine the technique of measurement so as to obtain an absolutely exact and unchanged value for the property being measured? Naturally, you will be hampered by the imperfection of the tools and of the human senses, but suppose you worked things out in principle, assuming the existence of perfect tools and perfect senses. Could you then obtain perfectly accurate values?

In order to do the best one possibly can, the tool used must be very small and delicate in comparison with the system whose properties are being measured. Thus, a tiny thermometer will absorb very little heat, and a tiny pressure gauge will lose very little air. The smaller the measuring tool, the less the measurement will be affected and, in principle, the more exact the measurement.

For an ultimately exact measurement, you must have an ultimately small measuring tool. If the ultimately small does not exist then neither does the ultimately exact.

The tiniest conceivable tools are, of course, the subatomic particles and they are small enough for anything short of ultimate exactness—or so it would seem.

However, as we descend into the subatomic world, we find ourselves trying to measure the properties of objects that are themselves extremely tiny. We must measure the properties of subatomic particles by using subatomic particles as tools. Our tools are then as large as the objects being measured and it is therefore inevitable that we have a great deal of trouble making exact measurements.

Consider an electron, for instance, and suppose we want to measure its momentum so that, eventually, we can tell whether the law of conservation of momentum holds exactly for the system of which it is part.

To do this we might spray photons in its general direction. Occasionally one photon will strike and rebound. From the direction in which the rebounding photon returns and from the time taken for it to go and return, we can determine the position of the electron at a particular instant of time. If this is done over and over, we will determine its position at different instants of time and from that fact calculate its velocity. If the velocity is obtained and if the mass of the electron is assumed to be known, then we have its momentum.

The only trouble is that the photon is likely to be as large as the electron and when it strikes the electron, the electron itself rebounds. The path it follows under the bombardment of photons is radically different from that it would have followed in the absence of the photons. Therefore, although you might have exact values of the position of the electrons at different instants of time, you would not have the faintest idea of what its velocity would have been in the absence of the photons.

You can try to get around this difficulty by using progressively less energetic photons—photons weak enough to introduce very little rebound in the electron. Then you might hope to determine its position while affecting the velocity less and less. In this way, the effect of a "zero-energy" photon might be calculated and the exact position and momentum of the electron simultaneously determined.

Unfortunately, the smaller the energy of a photon, the longer its wavelength; and the longer its wavelength, the less efficiently it bounces off the electron. It becomes more and more likely that it bends around the electron instead and bounces off, if at all, in the wrong direction. The result is that as one determines the momentum more accurately, it becomes more difficult to be sure of the position.

In 1927 Heisenberg analyzed the situation very carefully and was able to show that one could determine the momentum of any particle with as great an exactness as one chose; but that the more exactly the momentum was determined, the less exactly the position was determined. One could also determine the position as exactly as one chose, but the more exactly that was done, the less exactly one could determine the momentum.

In fact, Heisenberg showed that the inexactness of the momentum determination (which we may call the "uncertainty" of the momentum and symbolize as Δx). multiplied by the inexactness or uncertainty of the position (Δp) could never come to less than a certain fixed value in the case of any system at all, whether an electron or the Sun. He worked out the relationship as:

$$\Delta x \cdot \Delta p \geqq \frac{h}{2\pi}$$

where $\geqq$ represents "is equal to or more than," π (the Greek letter "pi") is the well-known constant equal to about 3.14159 and h is a quantity called *Planck's constant*. This equation is referred to as an expression of Heisenberg's *uncertainty principle*.

Planck's constant, originally derived by Planck in 1900 in connection with his quantum theory (see page 40) is an extremely small quantity. The value currently accepted is 0.000000000000000000000000000066256 or 6.6256×10^{-27} erg-seconds. The quantity $h/2\pi$ comes out then to almost exactly 10^{-27} erg-seconds. We can therefore write the equation for the uncertainty principle as follows:

$$\Delta x \cdot \Delta p \geqq 10^{-27}$$

The theoretical limit of certainty is thus small indeed. In theory—if one could obtain tools fine enough and senses sharp enough—one could, at the same moment of time, determine the position of an object to rather less than a trillionth of a centimeter of its "true" position, and its momentum to rather less than a trillionth of a gram-centimeter per second of its "true" momentum. Such refined measurements supply us with all we need (and indeed much more than we need) in our everyday world, or even in the ordinary microscopic world. Therein, we can be as "certain" as we need be.

In the world of the atom and of the subatomic particles, however, the uncertainty principle introduces a severe limitation to the accuracy of the data we can obtain, even in principle. A trillionth of a centimeter is a sizable length in the subatomic world, and if the electron is not positioned more sharply than that, it is a fuzzy object indeed. And if it is positioned more sharply, then the uncertainty in its momentum becomes correspondingly greater, and an uncertainty of a trillionth of a gm-cm/sec in the determination of an electron's momentum is already uncomfortably large.

The uncertainty principle bothered a number of physicists (including Albert Einstein) when it was first advanced because there was a certain displeasure at the thought that there was an unavoidable limit to the amount of knowledge we might gain about the universe. It was as though nature "weren't playing fair" if the uncertainty principle were correct, and that therefore the uncertainty principle must be wrong.

However, the uncertainty principle proved to be extremely useful. Theoretical physicists found that the only way they could explain the manner in which atoms absorbed and radiated energy was to suppose that particles always possessed wave properties—the lighter the particle, the more prominent the waves. The electron, which is a particularly light particle, must exhibit quite prominent wavelike properties and these were, indeed, detected in 1927, the very year in which the uncertainty principle was advanced. The fact that the electron was wavy meant at once that one couldn't speak of its exact position as though it were a tiny billiard ball—because it wasn't. The uncertainty principle which might have been a mere nuisance if the electron were a tiny billiard ball proved to be the only way by which an electron wave could be made to make sense.

Then, too, the uncertainty principle proved to be vital in working out the details of the nuclear field, and I will move in that direction, now, for this will bring us, in due course, back to the neutrino.

UNCERTAINTY AND CONSERVATION

In 1930, at a gathering of physicists at Brussels, Einstein tried to demonstrate a fallacy in the uncertainty principle. In this he failed. The reasoning he advanced to demonstrate the fallacy was itself shown to be fallacious by the Danish physicist Niels Bohr, and Einstein had to own himself beaten.

In the process of his argument, however, Einstein showed that if the uncertainty principle were correct, it could be expressed in terms of uncertainty of energy (Δe) multiplied by uncertainty in time (Δt), as well as in Heisenberg's fashion. We can then say:

$$\Delta e \cdot \Delta t \gtrsim 10^{-27}$$

This portion of Einstein's reasoning has been accepted, and it introduces an interesting possibility.

Einstein's version of the uncertainty principle indicates that the more closely we determine the energy content of a system, the less closely we can determine the actual instant of time at which the energy content has the determined value, and vice versa.

Under ordinary conditions, we determine the energy content of a system over some reasonable length of time, and therefore can, in principle, determine it with great exactness and convince ourselves that the law of the conservation of energy holds with equally great exactness.

But what if we were to attempt to determine the

energy content of a system over, say, a trillionth of a trillionth of a second? This means we must expect to determine the time with at least that great an exactness, and the uncertainty in energy content would then become quite large. We would have no way of telling whether the system would have the energy content it "ought" to have according to the law of conservation of energy. Because of the uncertainty in measurement, it might have considerably more or less than it "ought" to have.

Let's take an analogy. Suppose a schoolboy is strictly forbidden to show any impoliteness to his stern teacher at any time under pain of a severe flogging. How can the teacher tell that the boy is not sticking his tongue out everytime the teacher turns his back? The teacher may whirl about and find no outstretched tongue, but the teacher cannot turn in less than a definite time and the tongue can disappear faster than the teacher can turn.

You might say that it doesn't matter whether or not the teacher catches the boy. If the boy sticks his tongue out he has broken the rule, whether or not he is caught. And, by the same line of argument, the law of conservation of energy would demand that a system have a certain fixed quantity of energy, whether that fixed quantity can be measured exactly or not.

That may be true, but it is not the way the practical world of affairs runs. The practical rule for the schoolboy is not "Never be impolite," but "Never be caught being impolite." If the teacher does not catch that tongue out, he cannot, in good conscience, punish the boy. And if the energy content of a system cannot be measured exactly, we cannot, in good conscience, insist that the content must nevertheless be exactly so much.

In short, we must not define the law of conservation of energy as: "The total energy of a closed system remains constant," but "The total energy of a closed system remains constant within the limits of the uncertainty principle." Under this second, more reasonable definition, the law of conservation of energy in its absolute sense can be "broken" for a short period of time; the shorter the period of time, the more radically it can be "broken."

This flexible version of the law of conservation of energy can be used in working out the details of the nuclear field that was required to explain the existence of atomic nuclei of elements other than hydrogen.

The Japanese physicist Hideki Yukawa worked on the problem in the early 1930s and published his results in 1935. Essentially, he supposed that the nuclear field produced its strong attraction by means of an exchange particle.

This particle could be viewed as violating, by its very existence, the old pre-uncertainty form of the law of conservation of energy. This would mean it could exist only within the very short limit of time allowed it by the uncertainty principle.

Suppose a proton or neutron emits a particle which, under ordinary conditions, it lacks the energy to emit. Such a particle would have to be reabsorbed quickly, within the time limit set by the uncertainty principle. The emitted-and-absorbed particle cannot be detected by any conceivable device and it is called a *virtual particle*.

If such a virtual particle appeared within a nucleus and moved at the velocity of light it could travel the distance from one nucleon to another and back in about 0.000000000000000000000005 or 5×10^{-24} seconds. If this time lapse is set as the uncertainty in time (Δt), then from Einstein's version of the uncertainty principle we can calculate the uncertainty in energy content (Δe) of the proton emitting the virtual particle. This comes out to be about 0.0002 ergs, which is equivalent to about 125 Mev, which is in turn equal to a mass equal to about 250 times the mass of an electron.

To put it in another way: If a proton emitted a particle equal in mass to 250 times that of an electron, that particle could not be detected in less than 5×10^{-24} seconds. During that time interval, the proton is free to "break" the law of conservation of energy to the extent of 250 electron masses; and during that time interval the particle could get to the next nucleon and back if necessary.

If the virtual particle were considerably less massive than that, it would go undetected for a correspondingly longer period and might penetrate a perceptible distance beyond the nucleus. The nuclear field must then make itself evident outside the nucleus—which it doesn't. On the other hand, if the virtual particle were considerably more massive than 250 times that of an electron, it would not have time to reach the next nucleon and there would be no possibility of the nucleus hanging together.

In 1935, then, Yukawa predicted that the nucleus hung together by virtue of a nuclear field that expressed itself by means of the continual emission and absorption of particles with masses of about 250 times that of the electron. And the uncertainty principle explained why the nuclear field was as short-range as it was.

11

MUONS

TRAPPING THE MESON

Such exchange particles are all very well to talk about, but as long as they cannot be detected and their existence cannot in any way be made manifest, they must remain mere figments of theory.

But a virtual particle remains virtual only because the system producing it lacks the energy to make it real. If energy were added to the system, that energy could be converted into the mass of the particle, which would then exist outside the confines of the uncertainty principle. It could then, possibly, be detected.

For this purpose, though, at least 125 Mev of energy must be added to an atomic nucleus and in the early 1930s such energies could not easily be concentrated into atomic nuclei by means of the devices then available. The only source for such energies at the time were the extremely energetic cosmic rays bombarding Earth from outer space. Individual cosmic-ray particles may contain energies not merely in the hundreds of Mev, but in the billions of Mev. (Even today when physicists have built huge devices capable of producing streams of subatomic particles with energies of 30,000 Mev or more, the furious energies of the upper range of cosmic-ray particles remain far out of reach.)

It is now known that cosmic-ray particles consist of bare atomic nuclei that have been slowly accelerated on their vast journeys through interstellar space (probably by magnetic fields associated with stars and galaxies). Since the matter of the universe is very largely hydrogen, with helium making up most of what remains, it is not surprising that the cosmic-ray particles are about 78 percent protons (hydrogen nuclei), 20 percent alpha particles (helium nuclei), and 2 percent more massive nuclei.

These positively charged nuclei are the *primary radiation*. When these smash into the Earth's atmosphere, their powerful energies bring about a variety of changes in the nuclei they encounter. Very energetic particles are hurled out of these nuclei and these form the *secondary radiation*.

It would not be surprising if the secondary radiation consisted of speeding electrons and of high-energy photons, but some of the properties of this radiation did not match what was to be expected of electrons and photons.

Physicists investigating cosmic rays were therefore speculating during the early 1930s (quite apart from Yukawa's theory concerning the nuclear field) on the possible existence of particles more massive than the electron but less massive than the proton. Such particles of intermediate mass were needed to explain the data being gathered concerning cosmic rays.

In 1935, shortly after Yukawa's theory had been published, Anderson (who, three years earlier, had discovered the positron, see page 83) was investigating cosmic rays on Pikes Peak. The next year he studied the photographs he had obtained and found tracks that curved by an amount one would expect of a particle of intermediate mass, one which turned out to be about 207 times as massive as an electron.

Anderson called the particle a *mesotron,* from the Greek word "meso" meaning intermediate, but this was quickly abbreviated to *meson,* and it is this name which has grown prominent.

At first it was thought that Anderson's particle was Yukawa's exchange particle, brought into observable reality by the energy of cosmic rays, even though the particle was rather less massive than Yukawa had predicted.

Unfortunately the evidence argued against this. By the very nature of the nuclear field, it was necessary to suppose that Yukawa's exchange particles would interact very intensely and quickly with any nucleon it came across. It could not be expected, therefore, to penetrate far into matter since the first nucleus it approached would absorb it.

Anderson's particle, however, proved to penetrate matter easily, moving through sizable thicknesses of lead, for instance. It must, in the process, have encountered many nuclei without being absorbed. This meant it could not possibly be the nuclear exchange particle.

The disappointment was eased in 1948, however, thanks to a group of English physicists, headed by Cecil Frank Powell, who were studying cosmic rays from the vantage point of high altitudes in the Bolivian Andes. They detected particles distinctly more massive than Anderson's meson, particles possessing about 270 times the mass of an electron.

The mass of the new particle was closer to Yukawa's prediction and it interacted with matter with satisfactory intensity. It was Powell's meson and not Anderson's that proved to be the nuclear exchange particle. Yukawa's theory was established and so was the existence of the nuclear field.

Powell named his particle the *pi-meson*. The Greek letter "pi," equivalent to our "p," was adopted because it stood for "primary radiation" which Powell felt was responsible for the production of pi-mesons. Anderson's particle, the first meson to be discovered, had a sort of property-right to the initial letter "m," which in the form of the Greek letter "mu" was attached to its name. It became the *mu-meson*.

As time went on, additional types of mesons were discovered and it came to be realized that subatomic particles generally could be divided into three groups rather than two. There were not merely leptons and baryons, but mesons as well.

However, since I am on the track of the neutrino only, I will confine myself to a consideration of the pi-meson and mu-meson only.

STRONG AND WEAK INTERACTIONS

The establishment of the nuclear field did not immediately solve all problems. A puzzle arose in connection with the time-lapse required by meson interactions. A pi-meson flying past a nucleus at virtually the speed of light remains close enough to the nucleus to be within the sphere of influence of the very short-range nuclear field for only about 10^{-23} seconds at the most. In this infra-tiny bit of time, the pi-meson nevertheless has a chance to interact with the nucleus.

There seemed reason to think that all meson interactions ought to be equally speedy. In particular, one would expect pi-mesons and mu-mesons, when existing in the free state, to break down in no longer than 10^{-23} seconds.

But no; an isolated pi-meson, for instance, breaks down to less massive particles in about 0.0000000255 or 2.55×10^{-8} seconds. An isolated mu-meson is even longer-lived, breaking down to less massive particles in 0.000002212 or 2.212×10^{-6} seconds.

Such time intervals of tenths and hundredths of a millionth of a second seem extremely short to us, but on the subatomic scale they are actually excessively long. We can show this by making use of more familiar units of time.

Suppose we had a theory which led us to believe that a particle should break down in one second and instead we found that some endured for a hundred million years and others for ten billion years. We would be amazed at these tremendously long lives, would we not? Actually, the difference between the 10^{-23} seconds that the pi-meson was expected to last in theory and the 2.55×10^{-8} seconds that it endured in actual fact, is the same, proportionately, as the difference between one second and a hundred million years.

It was necessary to suppose, then, that there was no single nuclear field responsible for all meson interactions, but two nuclear fields, one much stronger than the other.

One field produces the *strong interactions,* such as the interaction of pi-mesons with nucleons, while the other produces the *weak interactions,* such as the breakdown of pi-mesons and mu-mesons. (It proved reasonable to include the breakdown of the neutron—see page 76—among the weak interactions as well.)

The pi-meson is the exchange particle for the strong interactions, and there should also be an exchange particle for the weak interactions. Fermi had worked out a theory of weak interactions which seemed to require such an exchange particle. This is sometimes referred to as the *w particle* (*w* for "weak," of course). The theory seems to indicate that in its free state, the *w* particle is a massive particle, more massive than a proton but will have a lifetime of only 10^{-17} seconds, which is about a billionth of the lifetime of a pi-meson, for instance. It is, therefore, not easy to detect.

There are now four different fields recognized by physicists—one or another of which is responsible for all the events that go on in the universe. These are the two nuclear fields, the electromagnetic field, and the gravitational field.

The strong nuclear field is the strongest of all, being rather more than a hundred times as strong as the electromagnetic field. The weak nuclear field is only a hundred-billionth as strong as the electromagnetic field but remains many trillions of times stronger than the gravitational field. Gravitation is still, as far as we know, the weakest force in nature by a goodly margin.*

---

*In 1964 physicists were driven by certain data, newly discovered, to consider the possibility of the existence of a fifth field, one which would be far weaker than even the gravitational field. Preliminary experiments in early 1965, however, made it seem that the fifth field did not exist.

THE MASSIVE ELECTRON

As the 1950s progressed, the mu-meson grew more and more puzzling. It fulfilled no essential role that physicists could reason out (not like the pi-meson which is essentially for nuclear stability). Furthermore, it slowly seemed to lose its identity and began to seem more and more like a variety of electron.

This may seem strange since the most prominent properties of the mu-meson seem to be quite different from those of the electron. The mu-meson is 207 times as massive as the electron, for one thing. For another, whereas the electron is a stable particle, the mu-meson is unstable, breaking down in 2.212×10^{-6} seconds.

Yet there are a number of properties which the mu-meson and the electron hold in common. We can list them:

1) The electron carries a charge of -1, and its anti-particle, the positron, carries a charge of $+1$. In this respect, the mu-meson is similar. It comes in two varieties, a *negative mu-meson* which, like the electron, carries a charge of -1 and is a particle, and a *positive mu-meson* which like the positron carries a charge of $+1$ and is an antiparticle. The negative mu-meson can be symbolized as μ^- (μ being the Greek letter "mu"). The positive mu-meson, being an antiparticle, can be symbolized as $\overline{\mu^+}$.

2) There is no "neutral electron," that is, no un-charged particle with the mass of an electron. There is no "neutral mu-meson" either.

3) The electron and positron have spins of $+\frac{1}{2}$ or $-\frac{1}{2}$, and this is true also for the negative and positive mu-meson.

4) The electron and positron are never involved in strong interactions but are involved in weak interactions; and this is true for the negative and positive mu-meson as well.

5) Finally, the magnetic properties of the electron and positron are virtually identical with those of the negative and positive mu-meson.

With all these similarities, how important are the differences in mass and stability?

As far as stability is concerned, that can be dismissed at once. I have already said that on the subatomic scale, a lifetime of 0.000002212 seconds is enormously long. Such a lifetime, compared to the time involved in strong interactions, is as ten billion years to 1 second. If we ourselves were used to events that took a second to transpire, we would be justified in considering an event that took ten billion years to transpire as "taking practically forever." In the same way we can say that on the subatomic scale, the mu-meson lasts "practically forever" and that the difference between its lifetime and the actually forever lifetime of the electron and positron is insignificant.

That leaves the mass difference as the great puzzle. Massive particles can be involved in both weak and strong interactions, while light particles, apparently, are involved only in weak interactions. The boundary line comes at the pi-meson; the pi-meson is the least massive particle known to be involved in strong interactions.

Yet the mu-meson, about three-quarters as massive as the pi-meson, doesn't make it. It falls short and is involved in weak interactions only. One might almost ask, "Why bother giving it all that mass, if not to make it capable of strong interactions?" There is, alas, no answer as yet to that question.

For that matter, why should the negative mu-meson be so like an electron despite the mass difference, and why should the positive mu-meson be so like the positron? And if the mu-mesons are indeed merely "massive electrons," why is their mass just 207 times that of an electron, no more and no less? Physicists do not have the answers to any of these questions as yet.

Still, though the answers elude us, we have no choice

but to consider the mu-mesons to be merely massive electrons and positrons, and in that case they must be considered leptons and be involved in the law of conservation of lepton number. The negative mu-meson, like the electron, is given a lepton number of +1, and the positive mu-meson, like the positron, is given a lepton number of −1. With these assignments, physicists have found that in all subatomic events involving the mu-meson, the law of conservation of lepton number is upheld.

Furthermore, since the mu-meson is a lepton, it seems misleading to continue calling it a "meson." The custom has therefore arisen of contracting the name "mu-meson" further, and referring to such particles as *muons*. There is, of course, the *negative muon* and the *positive muon*.

The pi-meson, on the other hand, deserves its name. For one thing it is neither a lepton nor a baryon. If the pi-meson is given a lepton number of 0 and a baryon number of 0, then in all subatomic events involving the pi-meson, the laws of conservation of lepton number and of baryon number are upheld.

Nevertheless, by the force of analogy with the muon, the pi-meson has come, more and more frequently, to be called the *pion*. The pion comes in two charged varieties, the *positive pion* (π^+) which, in this case, is the particle, and the *negative pion* ($\overline{\pi^-}$) which is the antiparticle.

Unlike the case of the electron and the muon, the pion also comes in an uncharged variety, the *neutral pion* (π^0) which is a trifle less massive than the charged pions, only 264 times as massive as an electron, and considerably shorter-lived, breaking down after only 1.9×10^{-16} seconds. Most unusual of all, the neutral pion, like the photon, is its own antiparticle.

If the muon is a massive version of the electron, it ought to be able to duplicate the function of an electron in atomic structure and this, indeed, it can do.

The electron is to be found in the outskirts of the

atoms where it can be pictured as a particle circling the atomic nucleus in certain specific orbits, or as a wave form occupying certain specific energy levels.

Negative muons can, under certain circumstances, briefly take the place of the electron in atoms. (And, presumably, positive muons can take the place of circling positrons in atoms of antimatter.) An atom in which a negative muon has replaced the electron is called a *mesonic atom*.

The difference in mass between the muon and electron introduces some changes of course. The angular momentum of a particle revolving about a nucleus depends upon the mass of the particle and its distance from the nucleus, among other things. Since the muon is 207 times the mass of the electron, it must decrease its distance from the nucleus correspondingly if angular momentum is to be conserved when the muon replaces the electron.

In very massive atoms, which ordinarily draw in their innermost electrons to close quarters indeed, a negative muon may circle the nucleus so closely as to be moving within that nucleus's outer perimeter. The fact that the muon can circle freely within the nucleus shows how small is the tendency of the muon to interact with protons or neutrons. (Here, too, the muon resembles the electron, which also has very little tendency to interact with nucleons. If that were not so, electrons would tend to be absorbed by nuclei and matter as we now know it would not exist.)

If the muon in the mesonic atom is pictured as a wave form at specific energy levels, then its greater mass requires that it occupy correspondingly higher energy levels than the electron, with correspondingly larger differences between adjacent levels. The photons emitted by the shift in energy level of a muon in a mesonic atom are again correspondingly larger, so that radiation emitted by mesonic atoms is uniformly in the X-ray region, whereas ordinary electronic atoms radiate in the visible region and in the ultraviolet.

To be sure, the mesonic atom is no more stable than the muon itself, for when the muon breaks down after a millionth of a second or so, the atomic nucleus will replace it with an ordinary electron.

12

THE MUON-NEUTRINO

PION BREAKDOWN

In particle interactions, the muon ought also to imitate the electron, if the muon is, indeed, merely a massive electron.

For instance, a negative pion will break down to form a negative muon, while a positive pion will break down to form a positive muon. This muon production should share the characteristics of electron production. Since an electron (or positron), when produced, is accompanied by an antineutrino (or neutrino), might not the same be expected when a muon is formed? Exactly this happens, in fact, so that we might write:

$$\pi^- \longrightarrow \mu^- + \bar{\nu}$$
$$\pi^+ \longrightarrow \overline{\mu^+} + \nu$$

In either case you will notice that the products of the breakdown have a net lepton number of 0. The law of conservation of lepton number requires that the particles before breakdown must therefore also have a lepton number of 0. The only patricles before breakdown are the negative and positive pion which, for this reason, must be assigned lepton numbers of 0. It would also appear from this interaction that there is no such thing as a "law of conservation of meson number," for when a pion breaks down to a muon, a meson disappears without replacement. Nor have physicists felt it necessary to adjust their theories to conserve mesons.

163

Things are quite comfortable, in that respect, as they are.

One might legitimately wonder, however, why the pion breaks down to a muon, if the muon is only a massive electron. Why does it not break down to an electron also? The answer is that, on occasion, it does.

In 1958 it was discovered that 1 pion out of 7000 will break down to form electrons rather than muons:

$$\overline{\pi^-} \longrightarrow e^- + \overline{v}$$
$$\pi^+ \longrightarrow \overline{e^+} + v$$

Why are muons and electrons not formed in equal numbers? There is a mass difference for one thing. Since a muon is so much more massive than an electron, the energy released on pion breakdown is tied up in mass formation to a considerable extent and comparatively little remains to be turned into kinetic energy. As a result, the muon, when formed, has a velocity of only about 25,000 miles per second. In forming the electron, very little energy need be tied up in mass and the electron speeds off at over 180,000 miles per second, very nearly the velocity of light.

Fermi, in working out the theory of weak interactions, showed that the likelihood of a muon, rather than an electron, being formed in the case of pion breakdown depends in part on the velocity of the particle formed. The closer the velocity to that of light, the less likely that particle is to be formed. It is for this reason that the slow muon is formed much more often than the rapid electron, in this particular case.

If variations due to mass difference are left out of account, it remains fair to say that for every particle interaction known to involve electrons (or positrons) there are analogous interactions involving negative muons (or positive muons).

And what about the neutrinos and antineutrinos formed along with electrons and muons? Are they identical?

At first, when the twinship of electrons and muons

was not appreciated, and when the muon was considered a distinct particle with properties not at all necessarily similar to the electron, there was no reason to think that the light, neutral particles formed in muon production were necessarily neutrinos.

Since the muon was so much more massive than the electron, it seemed only reasonable to suppose that the light, neutral particle formed along with it was more massive than the massless neutrino, though less massive, of course, than the neutron. For a while, then, physicists spoke of this neutral particle of supposedly intermediate size as the "neutretto." It was suspected of being actually more massive than the electron.

However, as the "neutretto" was studied more carefully, it was found that its mass had to be scaled downward again and again. More and more it began to look as though it were a massless neutral particle, like the neutrino. Then, when the essential identity of muon and electron was established, it became easy to suppose that neutrinos accompanied the formation of both electrons and muons, and that they were the same neutrinos in either case.

CONSERVATION OF ELECTRON FAMILY NUMBER AND OF MUON FAMILY NUMBER

If, however, the neutrino accompanying electron formation is identical with the neutrino accompanying muon formation, a new problem arises in connection with muon breakdown. A negative muon breaks down to form an electron, and a positive muon breaks down to form a positron. In the former case, an antineutrino ought to be formed also and in the latter case a neutrino, as follows:

$$\mu^- \longrightarrow e^- + \bar{v}$$
$$\mu^+ \longrightarrow e^+ + v$$

But now something goes wrong with lepton number. The negative muon has a lepton number of $+1$, while

the electron and antineutrino have lepton numbers of
+1 and —1, for a net value of 0. Again, the positive
muon has a lepton number of —1, while the positron
and neutrino have lepton numbers of —1 and +1, for
a net value of 0. Has the law of conservation of lepton
number been violated? Or must we assign muons lepton
numbers of 0?

Neither possibility is acceptable to nuclear physicists
for it would raise more problems than it would solve.
Instead, the situation is best resolved if a third particle
is considered as arising from a muon that is breaking
down. Suppose a negative muon, in breaking down,
gave rise not only to an electron and an antineutrino,
but to a neutrino as well; and suppose that a positive
muon gave rise to a positron and neutrino, and to an
antineutrino as well. Now the breakdowns could be
represented as:

$$\mu^- \longrightarrow e^- + \bar{v} + v$$
$$\mu^+ \longrightarrow e^+ + v + \bar{v}$$

Thus, if we start with a negative muon with a lepton
number of +1, we end with three particles with lepton
numbers of +1, —1, and +1, for a net of +1. If we
start with a positive muon with a lepton number of —1,
we end with three particles with lepton numbers of —1,
+1, and —1, for a net of —1. We have conserved
lepton number without depriving muons of their lepton
status.

But we are not in the clear. The presence of both a
neutrino and an antineutrino among the products of
muon breakdown brings up a new problem. It is generally true that a particle and antiparticle, when brought
to close quarters, can undergo mutual annihilation with
the production of photons of appropriate energy content.

It may be that neutrinos and antineutrinos are less
likely to undergo annihilation than are the ordinary
run of particles and antiparticles, but such annihilation
ought to take place at times, even if only rarely. If it
did, then there would be occasional instances when

a negative muon would break down to an electron plus photons and a positive muon would break down to a positron plus photons, and these photons would be easily detected. —Yet they never were, Why not?

One theory advanced to explain the nonappearance of photons involved the nonexistence of the w particle (see page 157). If the w particle did not exist there were reasons to suspect that the breakdown of muons to electrons and photons would proceed so excessively rarely that it might not be observed. However, the w particle was useful elsewhere in nuclear theory, and physicists searched for an alternate explanation.

One arose, as early as 1957, in the suggestion that the neutrino and antineutrino produced in muon breakdown were not true examples of particle and antiparticle! Perhaps the electron produces one kind of neutrino which we might call the *electron-neutrino* (with its twin the *electron-antineutrino*), while the muon produced another kind, the *muon-neutrino,* and of course the *muon-antineutrino*. The first set could be symbolized as v_e and $\overline{v_e}$, while the second set could be symbolized as v_μ and $\overline{v_\mu}$.

Consider muon breakdown in this new light. A negative muon breaks down to form an electron and an electron-antineutrino. The additional particle formed in the breakdown of the negative muon must therefore be a muon-neutrino. You then have an electron-antineutrino and a muon-neutrino which are not a particle/antiparticle combination and cannot be expected to undergo mutual annihilation.

A positive muon, by the same line of reasoning, breaks down to form a positron and an electron-neutrino* Also formed is a muon-antineutrino. We can write the breakdowns of the muon as follows now:

*Why not a "positron-neutrino"? Because the word "electron" here includes both electrons and positrons, just as the word "muon" includes both negative muons and positive muons. Here is one of the cases where it is most inconvenient to have the positron possess a separate name of its own instead of the proper name of "antielectron," or "positive electron."

$$\mu^- \longrightarrow e^- + \bar{\nu}_e + \nu\mu$$
$$\mu^+ \longrightarrow e^+ + \nu_e + \nu\mu$$

Under these conditions, lepton number is still conserved, but now the possibility arises of a pair of still tighter conservation laws. Suppose we subdivide the leptons into an *electron family* and a *muon family*.

Included in the electron family are the electron, the positron, the electron-neutrino, and the electron-antineutrino. The electron and electron-neutrino each have an *electron family number* of $+1$, while the positron and the electron-antineutrino each have one of -1.

Included in the muon family are the negative muon, the positive muon, the muon-neutrino, and the muon-antineutrino. The negative muon and the muon-neutrino would each have a *muon family number* of $+1$, while the positive muon and the muon-antineutrino would each have one of -1. (The remaining lepton, the photon, would be a member of neither family and would have an electron family number of 0 and a muon family number of 0. So, of course, would all mesons and baryons. Furthermore, particles of the electron family would have a muon family number of 0, and vice versa.)

The equations representing muon breakdown with both sorts of neutrinos produced illustrate the *law of conservation of electron family number* and the *law of conservation of muon family number* which hold, respectively, that *the net electron family number and the net muon family number of a closed system remains constant*.

We begin with a negative muon (muon family number, $+1$, and electron family number, 0). Three particles are produced: an electron, an electron-antineutrino, and a muon-neutrino. The muon family numbers are 0, 0, and $+1$, respectively, for a net of $+1$. The electron family numbers are $+1$, -1, and 0, respectively, for a net of 0. Thus, both muon family number and electron family number are conservative.

The same line of reasoning can be used for the break-down of the positive muon to show that there, too, both muon family number and electron family number are conserved.

We can show this to be true for some interactions involving electrons or muons earlier in the book, pro-vided we distinguish carefully among the neutrinos. The breakdown of a neutron is taken to involve the electron-antineutrino and can be written as:

$$n \longrightarrow p^+ + e^- + \overline{\nu}_e$$

If you check you will find that the electron family number is zero to begin with and to end with and that the same is true of the muon family number.

Again the breakdown of a negative pion can result in either a negative muon and a muon-antineutrino, or in an electron and an electron-antineutrino:

$$\overline{\pi}^- \longrightarrow \mu^- + \overline{\nu}\mu$$
$$\overline{\pi}^- \longrightarrow e^- + \overline{\nu}_e$$

A pion has a muon family number of 0 and an elec-tron family number of 0. In the first of the pion break-downs, the negative muon and the muon-antineutrino have muon family numbers of +1 and —1 for a net of 0. In the second, the electron and electron-antineutrino have electron family numbers of +1 and —1, respec-tively, for an electron family number of 0. The same reasoning will apply to the breakdown of the positive pion.

Indeed, physicists have discovered that in all particle interactions involving muons or electrons or both, muon family number and electron family number are sepa-rately conserved.

To be sure, the sum of the two (lepton number) is also conserved. However, it is clearly more important that two subdivisions are separately conserved than that of the sum of the two is conserved. For this reason, the conservation of lepton number has fallen into disuse, though it has never been violated, and physicists speak instead of the laws of conservation of electron family number and of muon family number.

THE TWO-NEUTRINO EXPERIMENT

Of course, these new conservation laws were only valid if the electron-neutrinos and the muon-neutrinos were, indeed, not identical in nature. Unfortunately, there seemed no property to which this lack of identity could be pinned. Both sets of neutrinos were massless and uncharged. Both had spins of $+\frac{1}{2}$ or $-\frac{1}{2}$ and both had an antiparticle twin. Where, then, lay the difference?

Physicists hesitated, therefore, to accept the postulated difference between the electron-neutrino and the muon-neutrino without additional evidence. They sought methods whereby they could plan an interaction which would go one way if the neutrinos were identical, and another way if they were not.

Such an experiment was devised and, in 1962, conducted in the laboratories at Brookhaven, Long Island. To conduct the experiment a beam of high-energy neutrinos was required. This was obtained by smashing high-energy protons into a beryllium target in order to produce an intense beam of pions, of both the positive and negative varieties (see Figure 13).

The pion beam traveled toward a wall of armor plate

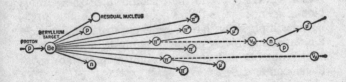

FIGURE 13. Production of Neutrinos in the Two-Neutrino Experiment

(from an old battleship) about 44 feet thick. Before the wall was reached, about ten percent of the highly unstable positive pions had broken down to positive muons and muon-neutrinos while an equal percentage of the negative pions had broken down to the negative muons and muon-antineutrinos.

The positive pions might also produce positrons and electron-neutrinos while the negative pions might also produce electrons and electron-antineutrinos, but in such small quantities (see page 164) that they might easily be neglected.

When the wall was struck by this conglomerate of particles, the pions and muons of both varieties were stopped. The muon-neutrinos and muon-antineutrinos kept right on going, however, passing through about 44 feet of armor plate as though it were a vacuum.

On the other side of the armor plate was an enclosure containing a ten-ton detection device, called a spark chamber, capable of reacting to certain nuclear events with great delicacy. Through this spark chamber there was passing a continuous stream of muon-neutrinos and muon-antineutrinos.

Very rarely, a muon-neutrino would react with a neutron to produce a proton and a negative muon. (At least, that was to be expected from theory.) The reaction can be represented as follows:

$$\nu\mu + n \longrightarrow p^+ + \mu^-$$

Here, as you see, baryon number is conserved, since a neutron is converted to a proton and both have a baryon number of $+1$. In addition, muon family number is conserved since a muon-neutrino is converted to a negative muon and both have a muon family number of $+1$.

This, at least, is what is to be expected if there were such a conservation law as that of muon family number. But what if there were not? What if the muon-neutrino were identical with the electron-neutrino and there were neither a muon family nor an electron family and only lepton number were conserved. In that case we would have to speak simply of a neutrino which on interaction with a neutron might produce a proton and a negative muon or a proton and an electron:

$$\nu + n \longrightarrow p^+ + \mu^-$$
$$\nu + n \longrightarrow p^+ + e^-$$

If there were only one kind of neutrino then theory would lead physicists to suppose, in this case at least, that there would be an equal probability of the formation of negative muons and of electrons and both would appear in equal numbers. In that case, only the conservation of lepton number could be used.

If there were two kinds of neutrinos, then, since only muon-neutrinos were entering the spark chamber, only negative muons should be formed. No electrons should be formed. In that case, there would be laws of conservation of electron family number and of muon family number.

By June 1962 about a hundred trillion neutrinos had passed through the spark chamber and had set off 51 "events." (There were also events produced by cosmic-ray particles and by other indirect causes—480 of them. These could be identified and discarded.)

Of the 51 neutrino-induced events, every one produced a negative muon; not one produced an electron. (The tracks produced by muons and electrons in spark chambers are quite different and can be easily distinguished.)

The conclusion derived from this "two-neutrino experiment" was that there were indeed two varieties of neutrinos, and that one could fairly speak of the laws of conservation of electron family number and of muon family number.

CONSERVATION OF PARITY

The conservation laws we have made use of so far in connection with the neutrino are seven in number.

They are the laws of conservation of:

1) momentum
2) angular momentum
3) energy
4) electric charge
5) baryon number
6) electron family number
7) muon family number

These are not the only conservation laws made use of by nuclear physicists, but with one exception they are all one needs for the neutrino story. The exception involves a quantity called *parity*, a strictly mathematical property which cannot be described in concrete terms. For our purposes it will be sufficient to say that to every particle can be assigned one of two varieties of parity: *odd parity* and *even parity*.

The usefulness of these terms rests in the fact that parity adds up exactly as do odd and even numbers in arithmetic. Two odd numbers always add up to an even number, for instance, and two even numbers also add up to an even number. For that reason, we can write:

Odd+Odd=Even
Even+Even=Even
Odd+Odd=Even+Even

Again, an odd number and an even number always add up to an odd number, so that we can write:

Odd+Even=Odd
Odd+Even=Odd+Even

In particle interactions, the same relationships seemed to hold. If a particle to which an odd-parity is assigned breaks down into two particles, those two particles proved to be one odd-parity and one even-parity. If a particle to which an even parity is assigned breaks down into two particles, both proved to be odd-parity, or both proved to be even-parity. No matter how complicated the situation, the requirements seemed to be just those one would find in odd and even numbers.

As long as these requirements were met one could

speak of the *law of conservation of parity* in which *the total parity of a closed system remained constant*.

Trouble, arose with the discovery of the *K-mesons* (sometimes called *kaons*) in the late 1940s. These are even more massive than pions, being some 966 times as massive as electrons—yet still only about half as massive as the neutron or proton.

K-mesons could break down in various ways. At times a K-meson would break down to two pions and at other times to three pions. Since a pion possesses odd-parity, two pions add up to even-parity, and three add up to odd-parity. (Three odd numbers always yield an odd sum.) In order to conserve parity, it was concluded that there were two kinds of K-meson, one with odd parity which broke down to three pions, and one with even parity which broke down to two pions. The two were distinguished by Greek-letter prefixes, the K-meson of odd-parity being called a "tau-meson," the other a "theta-meson."

And yet in every respect except parity, the two mesons were identical. Could parity alone and by itself suffice to differentiate between them, or could it be that the two particles were one, and that parity was not necessarily conserved?

In 1956 two Chinese-American physicists, Tsung-Dao Lee and Chen-Ning Yang, put forth theoretical reasons for suspecting that while parity might be conserved in strong interactions, it need not be conserved in weak interactions (and the breakdown of K-mesons is, of course, an example of a weak interaction).

There was now the necessity of working out some experiment which would go one way if parity were conserved and another way if it were not (as six years later, the same necessity arose in connection with distinguishing between electron-neutrinos and muon-neutrinos to see if the laws of conservation of electron family number and muon family number held).

The experimental method proposed arose out of a property called "handedness" (see Figure 14). This depends on the identity or nonidentity of an object with

OBJECT DIFFERENT FROM MIRROR IMAGE

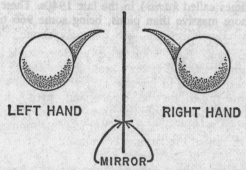

LEFT HAND   RIGHT HAND

MIRROR

OBJECT SAME AS MIRROR IMAGE

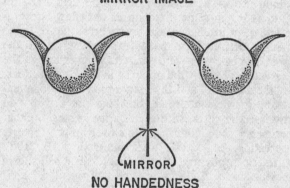

MIRROR

NO HANDEDNESS

FIGURE 14. "Handedness"

its mirror image. Your left hand, reflected in a mirror, looks like a right hand; your right hand, reflected in a mirror, looks like a left hand. Neither hand is identical with its mirror image because the existence of a thumb on only one side makes it asymmetrical. It is that which makes it possible to speak of a "left hand" and a "right hand." If your hand had a thumb on each side, it would be identical with its mirror image and it would be "nonhanded."

It can be shown that if parity is conserved, then subatomic particles act as though they are "nonhanded." There would be neither "right-hand particles" nor "left-hand particles." Such "nonhanded" particles would have to act symmetrically. If they broke down and emitted particles, those would have to be emitted in all directions equally. If, on the other hand, parity is not conserved, particles would be either left-handed or right-handed; the former would emit particles predominantly in one direction, the latter predominantly in the other.

Another Chinese-American physicist, Chien-Shung Wu, performed the necessary experiment. Atoms emitting beta particles (by means of a weak interaction) were cooled to near absolute zero and placed under the influence of a magnetic field. The field lined all the atoms in the same direction and the low temperature left them insufficient energy to break out of line.

Within forty-eight hours the experiment yielded the answer. The electrons were emitted asymmetrically. The conservation of parity did not exist in weak interactions. The "theta-meson" and the "tau-meson" were one and the same particle after all, breaking down with odd parity in some cases, with even parity in others. Other experiments soon confirmed the overthrow of parity conservation—at least in weak interactions. By 1959 the American physicist Maurice Goldhaber had shown that neutrinos and electrons were "left-handed," antineutrinos and positrons "right-handed."

And yet this brings up other problems. What is it that makes a particle "handed"? In particular, what makes a neutrino, which engages only in weak interactions, "left-handed" or "right-handed"? What is the source of this asymmetry in nature and why does it affect the weak interactions and not the strong?

As you see, then, the advances made by physicists result not only in solutions, but also in the opening up of new problems.

Why is the proton exactly 1836.11 times as massive as the electron? No one knows.

Why are there so many more electrons than positrons? No one knows.

Why is the muon 207 times as massive as the electron and just about identical otherwise? No one knows.

In what way are the muon-neutrino and electron-neutrino different, if both are identical in mass, charge, and spin? No one knows.

Why are particles symmetric in strong interactions and asymmetric in weak interactions? No one knows.

Nor am I regretful that I must end the story of the neutrino with puzzles. What would science be without puzzles to titillate and excite the scientist? And whence can we expect the great and thrilling discoveries to come if not out of those very same puzzles?

The riddle of the universe persists and always will, perhaps. The total and complete answer may never come. But with each generation the riddle moves to a higher and more sophisticated level and the game grows more rewarding and more delightful.

THE GRAVITON

Although I have come to what, at the moment, is the end of the story of the neutrino, there is a postscript.

You might think that nothing can be more ghostlike than the massless, chargeless neutrino, which can pass through light-years of solid matter without difficulty. And you might think that nothing can surpass the ingenuity of scientists who can predict the existence of such a particle and then not merely detect it, but detect it in four varieties.

But there is a particle more ghostlike than the neutrino.

There are four fields: 1) strong interactions, 2) electromagnetic interactions, 3) weak interactions, and 4) gravitational interactions. Of these the first three involve exchange forces and exchange particles. The exchange particle of the first is the pion, of the second the photon, of the third the w particle.

What of the fourth? What of the gravitational interaction? If it is a field, it too should have an exchange particle by means of which its influence is extended from one body to another. Since gravitation, like electromagnetism, is a long-range interaction, its exchange particle might be expected to be massless like the proton, rather than mass-possessing like the pion and the w particle.

The gravitational exchange particle, then, like the photon, ought to be a lepton. It ought to be electrically uncharged and ought to serve as its own antiparticle. In fact, the one clear way in which the gravitational exchange particle is different from the photon is in its spin. From theoretical considerations, physicists expect the gravitational exchange particle to have a spin of $+2$ or -2, where protons have spins of $+1$ and -1 (and where other leptons and baryons generally have spins of $+\frac{1}{2}$ and $-\frac{1}{2}$, while mesons have spins of 0).

Physicists have given the gravitational exchange particle a name—the *graviton*—and visualize bodies exerting their gravitational effects on each other by the constant emission and absorption of vast quantities these gravitons.

Gravitons have not yet been actually detected. For one thing, if they exist at all, they must be excessively low in energy. (This is to be expected considering the phenomenal weakness of the gravitational interaction. See page 140.)

In order to create high-energy gravitons which might be detectable, one would have to visualize a situation like that of a massive star moving back and forth rapidly and bodily. Stellar catastrophes do take place and stars explode or collapse violently, but in such cases, though huge masses accelerate outward or inward (or even both outward and inward alternately, as in the case of some variable stars) the center of gravity hardly budges. Parts of the star move rapidly, but the star as a whole does not move, and without bodily movement, the energy of the gravitons does not increase.

It may even be reasonable to suggest that here, at last, the theoretical physicist has hit an impassable barrier; that here at last he has a particle which is impossible to detect. However, in view of the accomplishments of science in the recent past, one does not dare make such a prediction. One dare not suggest that any feat is impossible.

Methods for detecting even the graviton may yet be devised. Let us wait and see.

APPENDIX

Exponential Numbers

In science the use of very large or very small numbers is unavoidable and the ordinary method of writing them is cumbersome. For instance, the Sun radiates over five octillion calories of energy per minute and this can be written as 5,600,000,000,000,000,000,000,000,000,000 cal/min. Or the mass of a proton is a little over a septillionth of a gram and this can be written as 0.000-000000000000000000000167252 grams. Numbers such as these, written in this fashion, are ponderous in appearance and are difficult to grasp and to work with.

Numbers can, however, be built up as multiples of tens. Thus, $10 \times 10 = 100$; $10 \times 10 \times 10 = 1000$; $10 \times 10 \times 10 \times 10 = 10,000$, and so on. Rather than write out the multiplication in full, the number of tens can be indicated by a small *exponent* written above and to the right of the figure 10.

We can write as 100 as 10^2 since it is 10×10; and 1000 can be written as 10^3, since it is $10 \times 10 \times 10$, and so on. An *exponential number* such as 10^x can be written as a 1 followed by x zeroes. Thus, 10^8, written out, would be 100,000,000 (a hundred million), 10^{12} would be 1,000,000,000,000 (a trillion), and so on.

Whenever we decrease the exponent by 1, the number itself is divided by 10. You see this when 10^4 (10,000) is reduced to 10^3 (1000) and when 10^3 (1000) is reduced to 10^2 (100).

Suppose we continue this. If we divide 100 by 10 we end with 10; and if we reduce the exponent of 10^2

(100) by 1, we end with 10^1. From this we must conclude that $10^1=10$.

If we reduce the exponent by one again, passing from 10^1 to 10^0, we divide the number itself by 10 again, from 10 to 1, so that $10^0=1$.

We can reduce 10^0 to 10^{-1} and in the process change the number 1 to 1/10, or 0.1. Therefore $10^{-1}=0.1$; and by the same line of argument $10^{-2}=0.01$, $10^{-3}=0.001$, and so on.

Negative exponents imply numbers less than one. It is best to write such numbers in decimal form with a single zero to the left of the decimal point as in the previous paragraph.

If this is done, then an exponential number of the form 10^{-x} is a decimal in which the number of zeroes (including the one to the left of the decimal point) is equal to x. Thus $10^{-6}=0.000001$ (a millionth) while $10^{-10}=0.0000000001$ (a ten-billionth).

What of numbers that are made up of digits other than 1 and 0? How would one express 6000 exponentially, for instance. We can convert 6000 into 6×1000. Since we know that $1000=10^3$, we can say that $6000=6\times10^3$.

In the same way $400,000=4\times10^5$ and $0.0000003=3\times10^{-7}$.

And if we had 6200 to worry about? We might write this as 62×100 and therefore as 62×10^2. There is nothing actually wrong with this but it is customary to keep the nonexponential portion of the figure between 1 and 10. Instead of writing 6200 as 62×100, we would write it as 6.2×1000. Therefore $6200=6.2\times10^3$.

In the same way 547 could be considered as 5.47×100 or 5.47×10^2. Again, $547,000,000=5.47\times100,000,000=5.47\times10^8$, and 0.0000478 is equal to 4.78×0.00001 or 4.78×10^{-5}.

If we go back to the radiation of the Sun, we find that 5,600,000,000,000,000,000,000,000,000 cal/min is equal to $5.6\times1,000,000,000,000,000,000,000,000,000$ or 5.6×10^{27} cal/min. Similarly, the mass of a proton, which is 0.00000000000000000000000167252 grams, is equal to $1.67252 \times 0.000000000000000000000001$ or 1.67252×10^{-24} grams.

Such numbers are not only more aesthetic in exponential form, but are much easier to handle in arithmetical manipulations in exponential form. They are therefore used quite generally by scientists and ought, these days, to be understood by anyone at all interested in science.

Alpha particles, 39, 51–52, 84
  atomic number and, 68
  cosmic rays and, 154
  energy of, 45–47, 97–101
  helium and, 43–44
  momentum and, 49–51
  penetration of, 98–100
  size of, 42
  spectrum of, 100
Alpha rays, 39
Aluminum, 44
Amber, 64
Anderson, Carl D., 83–84, 154–55
Angular momentum, 16–17
  conservation of, 16–19
  quanta and, 71–73
  solar system and, 31–33
Antideuteron, 95
Antielectron, 84
Antimatter, 94, 136–37
Antineutrino, 107
  detection of, 116–21
  Earth and, 123–24
Antineutron, 89–91
  breakdown of, 91–92
Antinucleons, 89–91
Antiparticles, 84
Antiproton, 88–89
Antiuniverse, 94–95
Argon, 127, 128
Aston, Francis W., 53

Atom(s)
  mass of, 41–45
  structure of, 47
Atomic nucleus, 43–44
Atomic number, 67–68
Atomic weight, 41–42

Baryon(s), 81, 92–94
Baryon number, 92–94
Becquerel, Antoine H., 39
Beryllium, 170
Beta particles, 39–40
  atomic number and, 68–69
  electrons and, 68–69
  energy of, 100–1
  formation of, 76–78
  natural production of, 122–23
  positive, 84–85
Beta rays, 39
Bethe, Hans A., 47–48
Bev, 46
Bohr, Niels, 149
Bragg, William H., 99

Cadmium chloride, 118
Calcium, 122
Calorie, 25
Carbon, 75
Carbon tetrachloride, 128
Cavendish, Henry, 140
Centimeter, 5n
Chadwick, James, 75, 103

Chamberlain, Owen, 88
Charge, electric, 64–68
Chemical reactions, 46–47, 58, 60
Chiu, Hong-Yee, 133–35
Chlorine, 127–28
Cloud chamber, 49–50
Coal, 33–34
Compton, Arthur H., 61
Condensation nuclei, 49
Cosmic rays, 83, 153–55
Coulomb, 66–67
Coulomb, Charles A. de, 66 138–39
Cowan, Jr., Clyde L., 117–18

Darwin, Charles R., 37
Davis, Raymond R., 128, 129
Deuterium, 95
Deuteron, 95
Dirac, Paul A. M., 71, 83
Du Fay, Charles F., 64–65
Dynes, 139

Earth, age of, 36–37, 48
    antineutrinos and, 122–24
Einstein, Albert, 53–55, 149
Electric charge, 64–65
    conservation of, 66
    units of, 66–67
Electromagnetic field, 142
Electromagnetic interactions, 139
Electron(s), 40
    atom and, 43
    atomic nuclei and, 76–78
    beta particles and, 68–69
    electric charge of, 66–67
    energy equivalence of, 60–61
    mass number of, 44

mu-meson and, 158–61
negative, 84
neutrino production from, 134
pion breakdown and, 163–64
positive, 83–84
positrons and, 86–88
spin of, 71–74
stability of, 82
symbol of, 68
Electron family number, 168
Electron-neutrino, 167
Electron-positron formation, 87–88
Electron-volt, 46
Electrostatic unit, 67
Energy
    conservation of, 22–27, 100–4, 150–51
    kinetic, 23
    manifestations of, 24
    mass and, 54–55
    nuclear, 47
    Sun and, 32–38
    units of, 24–26, 46n
Ergs, 25
Ev, 46
Evolution, 37
Exchange forces, 143
Exponent, 33n
Exponential numbers, 180–82

Fermi, Enrico, 103, 104, 157, 164
Fission, nuclear, 117
Franklin, Benjamin, 65–66

Galaxies, 136–37
Gamma rays, 39, 59
    energy of, 60–62

symbol of, 88
Gasoline, 55
Generalizations, 1–4, 27–28
Gev, 46
Goldhaber, Maurice, 176
Gram, 11
Gravitation, 157
    law of, 28
Gravitational constant, 140
Gravitational field, 142
Gravitational interactions, 139–40
Graviton, 178–79
Ground state, 99–100

Half-life, 79
Handedness, 174–76
Heat, 24–26
    mechanical equivalent of, 26
Heisenberg, Werner K., 75, 143, 147
Helium, 30, 39, 128
    atomic structure of, 43–44
    atomic weight of, 42
    fusion of, 131
    isotopes of, 44
    Sun and, 47–48
Helium-4, 51–52
    nuclear structure of, 75
Helmholtz, Herman L. von, 26, 35–36
Herschel, William, 28–29
Hydrogen, 42
    isotopes of, 44
    Sun and, 47–48, 124
Hydrogen-1, 52, 56, 70
Hydrogen-2, 94–95

Infrared radiation, 59, 60
Ions, 51
Iron, 132

Isotopes, 44–45
    atomic number of, 67–68
    symbols of, 45

Joliot-Curie, Frédéric, 84
Joliot-Curie, Irène, 84
Joule, James P., 26

Kaons, 174
Kev, 46
Kilocalorie, 25
Kilogram, 11
Kinetic energy, 23
Kirchhoff, Gustav R., 29
K-mesons, 174

Lavoisier, Antoine L., 21
Lead, 122
Lee, Tsung-Dao, 174
Lepton(s), 81–82
Lepton number, 107–9, 160
Light, 29–30
    quanta of, 40
    velocity of, 53–54
    wavelengths of, 58–60
Lockyer, Joseph N., 30

Magnesium, 131
Mass, 10
    conservation of, 20–22
    energy and, 54–55
    non-conservation of, 53
    nuclear reactions and, 51–53, 82
    velocity and, 62
Mass-energy, 56–58
Mass number, 44
Mass spectrograph, 53
Matter, 22
    degenerate, 132
Mayer, Julius R., 26
Measurement, 144–48
Mechanical forces, 139n
Meson, 154–55

Mesonic atoms, 161–62
Mesotron, 154–55
Meteorites, 34–35
Mev, 46
Microwaves, 59, 60
Millikan, Robert A., 67
Momentum, 11, 23
  angular, 16
  conservation of, 10–14,
    104–6
  linear, 16n
  unit of, 11n
Morrison, Philip, 133–34
Motion
  clockwise, 18–19
  component, 7–9
  counterclockwise, 18–19
  second law of, 3–4
Mu-meson, 155
  breakdown of, 156–57
  electron and, 158–61
Muon(s), 160
  atoms and, 160-62
  breakdown of, 165–68
  pion breakdown and,
    163–64
Muon family number, 168
Muon-neutrino, 167

Negative muons, 160, 161
Negative pions, 160
Neutral pions, 160
Neutretto, 165
Neutrino(s), 103
  absorption of, 113–16
  angular momentum and,
    106
  detection of, 127–30
  energy conservation and,
    100–3
  kinds of, 167–72
  momentum and, 104–6
  muon breakdown and,
    165–72

  particle spin and, 106
  Sun and, 124–26
  supernovas and, 132–35
  symbol of, 103–4
  universe and, 135–37
Neutrino astronomy, 122
Neutrino stars, 134–35
Neutron(s), 75
  breakdown of, 76–80
  magnetic field of, 90–91
Newton, Isaac, 3–4, 28, 29,
    53, 54, 139–40
Nitrogen, 71
Notation, exponential, 33n,
    180–82
Nuclear energy, 47
Nuclear field, 142–43
Nuclear interactions, 142
Nuclear reactions, 47
Nuclear spin, 73
Nucleons, 44
Nucleus, atomic, 43
  electric charge of, 67–68
  excited state of, 99
  positron formation and,
    84–85
  spin of, 73
  structure of, 43–45, 69–
    76, 138
Numbers, exponential, 33n,
    180–82

Oxygen, 75

Parity, 172–77
Particles, elementary, 81
  magnetic field of, 90–91
  massless, 63
  spin of, 71, 86–90
  stable, 82
  subatomic, 42
  uncharged, 73–75
  unstable, 82
  virtual, 151–52

Pauli, Wolfgang, 101–4
Perchloroethylene, 128
Phosphorus, 84, 85
Photons, 41
  absorption of, 111–13
  antimatter and, 95
  equivalent mass of, 58–63
  particle properties of, 61–62
  production by Sun of, 125
  spin of, 110
  stability of, 82
  velocity of, 62–63
Pi-mesons, 155
  breakdown of, 156–57
Pions, 160
  breakdown of, 163–64
Planck, Max, 40, 147
Planck's constant, 147
Polonium, 99n
Pontecorvo, Bruno, 127
Positive muons, 160
Positive pions, 160
Positron, 83–88
  stability of, 86
  symbol of, 84
Potassium-40, 122, 123
  half-life of, 79
Powell, Cecil F., 155
Primary radiation, 154
Protactinium, 69
Proton(s), 70–71
  cosmic rays and, 154
  mutual repulsion of, 138–39
  spin of, 72–73
  stability of, 91–94
Proton-antiproton formation, 88–89

Quanta(-um), 40–41

Radiations, radioactive, 39

Radioactivity, 39
Radio waves, 59, 60
Radium-226, 99
  half-life of, 99n
Radium-228, 52
Reines, Frederick, 117–18, 129
Relativity, theory of, 53–55, 62n, 63
Rest-mass, 62
Rutherford, Ernest, 42–43, 70, 79

Scalar quantity, 22
Scintillation counter, 98–99
Secondary radiation, 154
Segrè, Emilio, 88
Silicon, 84, 85
Solar constant, 32
Solar system, 30–32
Spectrum, 29
Spin, particle, 71–74, 106
Stars, 130–33
  binary, 28–29
  neutrinos and, 134–35
  second generation, 133
Strong interactions, 157
Subatomic particles, 42
Sulfur, 131–32
  atomic weight of, 42
Sun, 29–30
  center of, 125–26
  contraction of, 36
  energy of, 32–38, 47–48, 57–58
  hydrogen in, 124
  mass loss of, 57–58
  neutrino production in, 124–26
  temperature of, 125–26
Supernova, 132–33

Tau-meson, 174
Thales, 64

Theta-meson, 174

Thorium, 68–69

Thorium-*231*, 52

Thorium-*232*, 98
  breakdown of, 52, 122, 123
  half-life of, 79
  nuclear structure of, 75

Thorium-*234*, 51
  half-life of, 79

Time-travel, 137

Tin, 44, 45

Two-neutrino experiment, 170–72

Ultraviolet radiations, 59, 60

Uncertainty principle, 147–49

Universe, 135–37

Uranium, 39
  atomic weight of, 42
  energy in, 45–47
  isotopes of, 44, 75–76

Uranium-*235*, 52
  fission of, 117
  nuclear structure of, 75–76

Uranium-*238*, 51, 122, 123
  half-life of, 79
  nuclear structure of, 76

Vector quantity, 22

Velocity, 5, 23
  angular, 16

Virtual particle, 151–52

Volt, 46

Volta, Alessandro, 46

Wallis, John, 12

Weak interactions, 157

White dwarf, 133

W particle, 157, 167

Wu, Chien-Shung, 176

X rays, 59, 60

Yang, Chen-Ning, 174

Yukawa, Hideki, 151–52

DISCUS BOOKS
DISTINGUISHED NONFICTION

THEATER, FILM, AND TELEVISION

ACTORS TALK ABOUT ACTING
Lewis Funke and John Booth, Eds. 15062 1.95
ACTION FOR CHILDREN'S TELEVISION 10090 1.25
ANTONIN ARTAUD Bettina L. Knapp 12062 1.65
A BOOK ON THE OPEN THEATER Robert Pasoli 12047 1.65
THE CONCISE ENCYCLOPEDIC GUIDE TO SHAKESPEARE
Michael Martin and Richard Harrier, Eds. 16832 2.65
THE DISNEY VERSION Richard Schickel 08953 1.25
EDWARD ALBEE: A PLAYWRIGHT IN PROTEST
Michael E. Rutenberg 11916 1.65
THE EMPTY SPACE Peter Brook 19802 1.65
EXPERIMENTAL THEATER James Roose-Evans 11981 1.65
FOUR CENTURIES OF SHAKESPEARIAN CRITICISM
Frank Kermode, Ed. 20131 1.95
GUERILLA STREET THEATRE Henry Lesnick, Ed. 15198 2.45
THE HOLLYWOOD SCREENWRITERS
Richard Corliss 12450 1.95
**IN SEARCH OF LIGHT: THE BROADCASTS OF
EDWARD R. MURROW** Edward Bliss, Ed. 19372 1.95
INTERVIEWS WITH FILM DIRECTORS
Andrew Sarris 21568 1.65
MOVIES FOR KIDS Edith Zornow and Ruth Goldstein 17012 1.65
PICTURE Lillian Ross 08839 1.25
THE LIVING THEATRE Pierre Biner 17640 1.65
PUBLIC DOMAIN Richard Schechner 12104 1.65
RADICAL THEATRE NOTEBOOK Arthur Sainer 22442 2.65

GENERAL NON-FICTION

ADDING A DIMENSION Isaac Asimov 22673 1.25
A TESTAMENT Frank Lloyd Wright 12039 1.65
THE AMERICAN CHALLENGE
J. J. Servan Schreiber 11965 1.65
AMERICA THE RAPED Gene Marine 09373 1.25
ARE YOU RUNNING WITH ME, JESUS?
Malcolm Boyd 09993 1.25
BLACK HISTORY: LOST, STOLEN, OR STRAYED
Otto Lindenmeyer 09167 1.25
THE BOOK OF IMAGINARY BEINGS
Jorge Luis Borges 11080 1.45
BUILDING THE EARTH Pierre de Chardin 08938 1.25
CHEYENNE AUTUMN Mari Sandoz 09001 1.25
THE CHILD IN THE FAMILY Maria Montessori 09571 1.25
CHINA: SCIENCE WALKS ON TWO LEGS
Science for the People 20123 1.75
CLASSICS REVISITED Kenneth Rexroth 08920 1.25
THE COMPLETE HOME MEDICAL ENCYCLOPEDIA
Dr. Harold T. Hyman 15214 1.95

DISCUS BOOKS
DISTINGUISHED NON-FICTION

THE CONCISE ENCYCLOPEDIC GUIDE TO SHAKESPEARE Michael Rheta Martin and Richard A. Harrier	16832	2.65
CONSCIOUSNESS AND REALITY Charles Musees and Arthur M. Young, Eds.	18903	2.45
CONVERSATIONS WITH JORGE LUIS BORGES Richard Burgin	11908	1.65
DIALOGUE WITH SAMMY J. McDougall and S. Libovici	15644	1.65
DISINHERITED Dale Van Every	09555	1.25
DIVISION STREET: AMERICA Studs Terkel	22780	2.25
DRUG AWARENESS Richard E. Horman and Allan M. Fox (editors)	11064	1.45
ESCAPE FROM FREEDOM Erich Fromm	15727	1.95
FRONTIERS OF CONSCIOUSNESS John White, ed.	24810	2.50
GERMANS George Bailey	20644	1.95
GERTRUDE STEIN: A COMPOSITE PORTRAIT Linda Simon, Ed.	20115	1.65
GOODBYE, JEHOVAH William R. Miller	11163	1.45
THE GREAT ENGLISH AND AMERICAN ESSAYS Edmund Fuller	02311	.75
THE GREAT POLITICAL THEORIES, VOL. I Michael Curtis	23119	1.95
THE GREAT POLITICAL THEORIES, VOL. II Michael Curtis	23127	1.95
THE GREEK WAY Edith Hamilton	14225	1.50
GROTOWSKI Raymond Temkine	12278	1.65
HARD TIMES Studs Terkel	22798	2.25
HOMOSEXUAL: LIBERATION AND OPPRESSION Dennis Altman	14214	1.65
HOW OLD WILL YOU BE IN 1984? ed. by Diane Divoky	09142	1.25
THE HUMAN USE OF HUMAN BEINGS Norbert Wiener	21584	1.95
THE INCAS Garcilaso de la Vega	11999	1.65
INTERPRETATION OF DREAMS Freud	12393	1.95
JESUS IN BAD COMPANY Adolf Holl	19281	1.65
JUSTICE Richard Harris	12120	1.65
THE LIFE AND DEATH OF LENIN Robert Payne	12161	1.65
THE LIFE AND WORK OF WILHELM REICH M. Cattier	14928	1.65
LIFE IN A CRYSTAL PALACE Alan Harrington	15784	1.65
THE LIFE OF JOHN MAYNARD KEYNES R. F. Harrod	12625	2.45
MALE AND FEMALE UNDER 18 Nancy Larrick and Eve Merriam, Eds.	17624	1.25
A MALE GUIDE TO WOMEN'S LIBERATION Gene Marine	18291	1.65
MAN'S RISE TO CIVILIZATION P. Farb	21576	1.95
MAN IN THE TRAP Elsworth F. Baker, Ph.D.	18309	1.95
THE ME NOBODY KNOWS ed. Stephen M. Joseph	05264	.95
THE NATURE OF HUMAN NATURE Alex Comfort	08797	1.25

DISCUS BOOKS

DISTINGUISHED NON-FICTION

NATURE OF POLITICS M. Curtis	12401	1.95
THE NEW GROUP THERAPIES Hendrick M. Ruitenbeek	09647	1.25
NOTES OF A PROCESSED BROTHER Donald Reeves	14175	1.95
OF TIME AND SPACE AND OTHER THINGS Isaac Asimov	24166	1.50
THE OMNI-AMERICANS Albert Murray	11460	1.50
ON CONTEMPORARY LITERATURE Richard Kostelanetz (editor)	12385	1.95
THE PARIS AND NEW YORK DIARIES OF NED ROREM Ned Rorem	12617	2.45
POLITICS AND THE NOVEL Irving Howe	11932	1.65
THE POWER TACTICS OF JESUS CHRIST AND OTHER ESSAYS Jay Haley	11924	1.65
PRISONERS OF PSYCHIATRY Bruce Ennis	19299	1.65
THE PSYCHOANALYTIC REVOLUTION Marthe Robert	08763	1.25
THE LIFE OF EZRA POUND Noel Stock	20909	2.65
THE QUIET CRISIS Stewart Udall	24406	1.75
THE ROMAN WAY Edith Hamilton	14233	1.50
RUSSIA AT WAR Alexander Werth	12070	1.65
THE SCHOOLCHILDREN: GROWING UP IN THE SLUMS Mary Frances Greene and Orletta Ryan	18929	1.65
STUDIES ON HYSTERIA Freud and Breuer	16923	1.95
THE TALES OF RABBI NACHMAN Martin Buber	11106	1.45
THINKING ABOUT THE UNTHINKABLE Herman Kahn	12013	1.65
THINKING IS CHILD'S PLAY Evelyn Sharp	11072	1.45
THOMAS WOODROW WILSON Freud and Bullitt	08680	1.25
THREE NEGRO CLASSICS Introduction by John Hope Franklin	16931	1.65
THREE ESSAYS ON THE THEORY OF SEXUALITY Sigmund Freud	11957	1.65
TOWARDS A VISUAL CULTURE Caleb Gattegno	11940	1.65
THE WAR BUSINESS George Thayer	09308	1.25
WHAT WE OWE CHILDREN Caleb Gattegno	12005	1.65
WHEN YOU SEE THIS, REMEMBER ME: GERTRUDE STEIN IN PERSON W. G. Rogers	15610	1.65
WILHELM REICH: A PERSONAL BIOGRAPHY I. O. Reich	12138	1.65
WOMEN'S ROLE IN CONTEMPORARY SOCIETY	12641	2.45
WRITERS ON THE LEFT Daniel Aaron	12187	1.65

Wherever better paperbacks are sold, or directly from the publisher. Include 25¢ per copy for mailing; allow three weeks for delivery.

Avon Books, Mail Order Dept.
250 West 55th Street, New York, N. Y. 10019

DDB 7/75